ORDER PARAMETERS AND DOMAIN TOPOLOGY IN MAGNETICALLY INDUCED FERROELECTRICS

Dissertation

zur

Erlangung des Doktorgrades (Dr. rer. nat.)

der

Mathematisch-Naturwissenschaftlichen Fakultät

der

Rheinischen Friedrich-Wilhelms-Universität Bonn

vorgelegt von

Dennis Meier

aus Neustadt in Holstein

Bonn, November 25, 2009

Bibliografische Information der Deutschen Nationalbibliothek

Die Deutsche Nationalbibliothek verzeichnet diese Publikation in der
Deutschen Nationalbibliografie; detaillierte bibliografische Daten sind
im Internet über http://dnb.d-nb.de abrufbar.

Titelbild: Steffen Broßeit (BrossBoss Entertainment)

ISBN 978-3-8325-2489-0

Logos Verlag Berlin GmbH
Comeniushof, Gubener Str. 47,
10243 Berlin
Tel.: +49 (0)30 42 85 10 90
Fax: +49 (0)30 42 85 10 92
INTERNET: http://www.logos-verlag.de

Angefertigt mit Genehmigung der Mathematisch-
Naturwissenschaftlichen Fakultät der Rheinischen Friedrich-
Wilhelms-Universität Bonn

1. Gutachter: Prof. Dr. Manfred Fiebig
2. Gutachter: Prof. Dr. Markus Braden
3. Gutachter: Prof. Dr. Nicola Spaldin

Tag der Abgabe: 25. November 2009
Tag der Promotion: 16. April 2010

Abstract

In the field of strongly correlated electron systems, materials with coexisting magnetic and electric order are intensely discussed. These so-called multiferroics potentially host pronounced magnetoelectric interactions between the magnetic and the electric subsystem that enable magnetic phase control by an electric voltage – a highly desirable property for the design of multifunctional spintronics devices. Among them, systems where a spontaneous polarization forms as a direct consequence of magnetic long-range order are of particular interest. They naturally exhibit intrinsically strong and robust magnetoelectric interactions.

This fascinating class of materials, namely "magnetically induced ferroelectrics", is studied in the framework of this thesis by optical second harmonic generation (SHG). Special attention is paid to the analysis of the multiferroic domains in magnetically induced ferroelectrics, because they play an essential role with respect to long-term applications. The domains determine the switching of information bits in memory devices and the technological performance of permanent magnets. Moreover, at its root, any magnetoelectric interaction in a multiferroic corresponds to an interaction of its magnetic and electric domains.

SHG spectroscopy and temperature-dependent measurements on two selected model compounds ($MnWO_4$ and $TbMn_2O_5$) reveal that the symmetry-breaking magnetic and electric order parameters that drive the multiferroicity in magnetically induced ferroelectrics are inseparably entangled. This entanglement directly reflects the unique interplay of magnetism and ferroelectricity, clearly separating these systems from other multiferroics where magnetic order and ferroelectricity emerge independently.

For the first time, the inherent domain structure of a magnetically induced ferroelectric is visualized – the full three-dimensional distribution is resolved.

Hereby, two fundamentally different types of multiferroic domains are identified, denoted as magnetic translation and "hybrid-multiferroic" domains. The latter ones unify features that are associated to a magnetic domain state and others that point unambiguously to ferroelectric domains. Hence, a description in terms of ferroelectric or antiferromagnetic domains is incomplete and no longer appropriate.

Thus, this works evidences that the consequences of the rigid magnetoelectric coupling manifest also on the level of the domain structure and, therefore, can be exploited in future devices. Moreover, spatially-resolved measurements of the domain topology reveal that an electric voltage can be applied to uniquely control the multiferroic domain state. Remarkably, even after long-term tests no aspects of fatigue are observed.

In addition to the intriguing dynamic properties, a topological memory effect in $MnWO_4$ is discovered that allows one to reconstruct the entire multiferroic multidomain structure subsequent to quenching it.

Contents

Chapter 1

Introduction

One of the most intensively studied areas of research in condensed-matter physics is the field of strongly correlated electron materials [1, 2]. Typical examples for systems with strongly correlated electrons are found among the complex transition-metal oxides. These compounds are often formed by rather simple building blocks of transition-metal ions (elements that have an incompletely filled d sub-shell) and oxygen ions. Nevertheless, highly nontrivial phenomena are observed in this class of materials.

The complexity of strongly correlated electron materials is due to interactions between the electronic spins, charge, and orbitals that induce a variety of electronic phases. It is the competition and/or cooperation among these electronic phases that lead to fascinating physical effects.

Special highlights in this context are the discovery of high-temperature superconductivity in $YBa_2Cu_3O_{7-\delta}$ by Bednorz and Müller [3] or the verification of a giant magnetoresistance (GMR) in structures composed of alternating ferromagnetic and nonmagnetic layers by Grüneberg [4] and Fert [5]. Another important example is the observation of colossal magnetoresistance (CMR) in manganite thin films ($La_{2/3}Ba_{1/3}MnO_x$ [6] or $La_{0.72}Ca_{0.25}MnO_x$ [7]).

Only recently a whole class of strongly correlated materials, in which magnetic and electric order coexist, attracted tremendous attention. After Kimura et al. and Hur et al. revealed these so-called "multiferroics" to potentially host a gigantic magnetoelectric effect [8, 9], i.e. a coupling between the magnetic and the electric subsystem, numerous publications on spectacular correlation effects in multiferroic materials followed [10–15].

The broad current interest in multiferroic materials with pronounced magnetoelectric interactions is mainly triggered by their promising properties with respect to multifunctional spintronic devices [16]. Nowadays, ferroelectric materials are widely used in the sensor industry or to design ferroelectric random access memories (FERAM) in which information is stored by the remanent polarization [17]. Alternatively, ferromagnets are applied for long time data storage or magnetic field sensors. Multiferroics should allow for combining the advantages of both materials, enabling ferroelectric memories with non-destructive magnetic reading or magnetic random access memories (MRAM) with an electric writing procedure, called MERAM [18–20]. Especially the possibility of a magnetic phase control by application of an electric voltage could speed-up the switching of magnetic information bits and reduce the power needed for the switching process which would allow for further minimization of technical devices [21].

However, although the principle functionality of different devices based on multiferroic $BiFeO_3$ has already been demonstrated at room temperature, multiferroics have not entered

mainstream products as computers yet [22,23]. This partially follows from the fact that
the strongest magnetoelectric interactions in multiferroic materials are usually observed at
low temperature being unfeasible for electronic devices [15,24–27].

Nevertheless, regarding long-term applications some specific classes of multiferroics are
particularly promising. Extraordinarily strong interactions between magnetism and electric order are for instance common for multiferroics where a spontaneous polarization forms
as a direct consequence of magnetic long-range order [28,29]. In these so-called magnetically induced ferroelectrics the ordered spins violate the spatial inversion symmetry and
induce a ferroelectric polarization [30,31]. Although the polarization P_{sp} is typically small
($P_{sp} \propto 1 - 10$ nC/cm^2) compared to many established ferroelectrics ($P_{sp} \geq 10$ μC/cm^2),
it is the outstanding intrinsic robustness of the magnetoelectric coupling that renders
these systems interesting with respect to future applications. Remarkably, magnetically
induced ferroelectricity naturally emerges in CuO at temperatures up to 230 K [32]. Thus,
the underlying mechanism is particularly promising and may even provide the highly-
desired strong correlation between magnetic and electronic degrees of freedom at room
temperature which explains the tremendous activity in the field of magnetically induced
ferroelectrics [33–37].

In spite of the current interest in these compounds essential aspects of their multiferroicity
and the intrinsically strong magnetoelectric interactions continue to be a mystery. Moreover, nearly nothing is known about their domain topology which is an essential feature
of any ferroic system. It is the physics of the domains that determines the switching of
information bits in the abovementioned memory devices or the technological performance
of permanent magnets. At its roots, any magnetoelectric interaction in multiferroics originates from the correlation of magnetic and electric domains. Thus, understanding giant
magnetoelectric effects means understanding the domain states in this class of multiferroics [38].

Although domains are known to be present in magnetically induced ferroelectrics, they
cannot be imaged in a straight forward way. Often an antiferromagnetic order of the
spins is observed so that no macroscopic magnetic moments exist that can be exploited
to image the magnetic domain structure, whereas the small value of the polarization prohibits application of established imaging techniques such as piezoforce response microscopy
(PFM) for mapping the ferroelectric domains. Therefore, most previous studies focussed
on "removing" the magnetic/electric domains by converting the sample to or in between
single-domain states in (mostly indirect) experiments [25,26,32,39–41].

The aim of this thesis is to further the fundamental understanding of the complex interplay
of magnetism and electric order in magnetically induced ferroelectrics. A nonlinear optical
method, namely "second harmonic generation" (SHG) has been applied to investigate the
evolution of magnetic and ferroelectric order in two representative model compounds for
this class of materials, namely MnWO$_4$ and TbMn$_2$O$_5$. These systems were chosen as
model end compounds with respect to their very different micrsocopy and distinct spin-
spin interactions participating in driving the multiferroic state. Due to the differences the
observation of analogies regarding their physical properties will strongly point towards a
common feature considering the whole class of materials.

Optical SHG is a powerful technique for studying the multiferroic properties of MnWO$_4$
and TbMn$_2$O$_5$, because it is sensitive to the symmetry reduction imposed by the ordered
spins and the spontaneous polarization. Thus, SHG measurements simultaneously provide
access to both the magnetic *and* the electric subsystem which is an enormous advantage
regarding a study of the intricate correlation effects in magnetically induced ferroelectrics.

Furthermore, the spatial degree of freedom of SHG measurements allows for imaging of the magnetic and electric domain topology. By SHG topography the inherent domain structure of $MnWO_4$ and, in particular, its response to thermal annealing cycles and external fields has been revealed.

The thesis is organized as follows:

Chapter 2 provides the mathematical background for a later discussion and description of magnetic and electric order on the basis of the Landau theory of phase transitions and general group theory. Important terms and definitions are introduced. Special emphasis is put on the introduction of phases that exhibit a periodically modulated order, being an essential property of many magnetically induced ferroelectrics.

The **third chapter** is devoted to multiferroics and the basic mechanisms that are known to cause multiferroicity in crystals. After distinguishing two fundamental classes of multiferroic materials, different models are reviewed that explain the formation of a spontaneous polarization as a consequence of magnetic order. Moreover, the general concept of magnetic/electric domains is briefly outlined.

Optical second harmonic generation and applied experimental methods are discussed in **chapter 4**. A rather detailed description is provided regarding domain topography by SHG, because, in the context of this thesis, this technique has been applied for the first time to image the complex domain stucture in systems with periodically modulated antiferromagnetic order and, in particular, in magnetically induced ferroelectrics.

In **chapter 5** the relevant physical properties of the two chosen model systems $MnWO_4$ and $TbMn_2O_5$ are reviewed. The symmetry-breaking order parameters and point symmetries of the different ordered phases are discussed, being the basis for the derivation of symmetry-allowed SHG contributions in the two compounds. All symmetry-allowed tensor components of the nonlinear susceptibility are listed in this chapter.

The knowledge about nonzero tensor components is a prerequisite for the analysis of the experimental results presented in chapters 6 and 7. While in **chapter 6** polarization-dependent SHG spectra and temperature-dependent measurements are shown and interpreted with respect to the evolution of the magnetic and correlated electric order in $MnWO_4$ and $TbMn_2O_5$, **chapter 7** deals with the domain topology in magnetically induced ferroelectrics and its manipulation.

Chapter 2

Mathematical framework

This chapter provides the mathematical background for the later analysis and description of structural phase transitions and ordered phases in crystals. It is devoted to reviewing relevant principles of the Landau theory. Special emphasis is put on the description of phases that lack three-dimensional translation symmetry, called incommensurate phases.

2.1 Landau theory of phase transitions

The Landau theory is a phenomenological theory that provides a convenient formalism for the description of structural phase transitions. Instead of analyzing the microscopy, the Landau theory considers the general properties of a system close to a phase transition. Thus, it is valid irrespectively of the microscopic details of the material.

2.1.1 Ferroic phase transitions

The classical Landau theory applies to so-called *ferroic phase transitions*. Ferroic phase transitions are defined as nondisruptive phase transitions, that involve a change of the point-group symmetry of a crystal [42].

Here, the nondisruption condition is of fundamental importance. Nondisruptive basically means that the crystal structures above and below the transition can be transformed into each other in a continuous way [43]. In this case one can speak of conserved or violated symmetry operations in the context of a transition and consequences of the phase transition derived from symmetry arguments hold true.

An example for a nondisruptive transition is a structural transition across which a group-subgroup relation [44] is obeyed and a continuous deformation connects the high-temperature and the low-temperature structure. Another example are magnetic transitions at which the crystallographic structure is unaffected, whereas a continuous reorientation of spins is observed.

In contrast, the nondisruption condition is usually violated at martensitic transitions (also called shear or displacive transitions). All martensitic hcp↔bcc transitions for instance are *disruptive* or *reconstructive* phase transitions [42, 45]. In the following we generally assume that the nondisruption condition is respected.

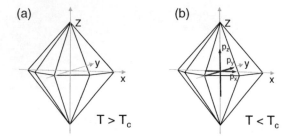

Figure 2.1: (a) Ditetragonal pyramid with point symmetry $4/mmm$. (b) For $T < T_c$ the ferroelectric order parameter can either lie within the xy plain (p_x, p_y) or point along the z direction (p_z).

2.1.2   Free energy and order parameters

We now discuss the standard procedure that allows to determine an expression for the free energy of a system. As an example a ferroelectric phase transition in a tetragonal crystal is considered [46].

In its high-temperature paraelectric phase (phase I for $T > T_c$) the crystal structure is assumed to have point symmetry $\hat{G}_0 = 4/mmm$ (time inversion symmetry is neglected). The spontaneous polarization $\mathbf{p}^{\mathrm{sp}} = (p_x, p_y, p_z)$ in phase I is zero and the structure remains unchanged under application of the 16 symmetry operations belonging to $\hat{G}_0$:

$$1, \; \bar{1}, \; 2_x, \; 2_y, \; 2_z, \; 2_{xy}, \; 2_{-xy}, \; \bar{2}_x, \; \bar{2}_y, \; \bar{2}_z, \; \bar{2}_{xy}, \; \bar{2}_{-xy}, \; \pm 4_z, \; \pm \bar{4}_z \,. \tag{2.1}$$

Here, the subscripts indicate the direction of the rotation- or rotation-inversion axis [47]. On example for a body with point symmetry $4/mmm$ is given in figure 2.1(a), showing a ditetragonal pyramid.

Since the set of symmetry operations $(\pm 4_z, \bar{2}_x, \bar{2}_z)$ allows to construct all aforementioned 16 elements of $\hat{G}_0$, the three symmetry operations are called *generators* of the point group. The inversion operation for example can be represented by $\bar{1} = 4_z \circ 4_z \circ \bar{2}_z$. Note that the choice of generators is not unique for a given point group.

For $T < T_c$ a ferroelectric phase (II) with $\mathbf{p}^{\mathrm{sp}} \neq 0$ becomes stable. As a consequence, for instance, inversion symmetry cannot be preserved, due to the polar displacement implied by $\mathbf{p}^{\mathrm{sp}}$ and thus the point group changes with respect to phase I. Here we assume a ferroic transition with a group-subgroup relation, so that the point group $\hat{G}$ of phase II is a subgroup of $\hat{G}_0$, i.e. $\hat{G} \subset \hat{G}_0$.

The basic idea of the Landau theory is to consider the polar displacement, specified by $\mathbf{p}^{\mathrm{sp}}$, as a variational degree of freedom. At all temperatures T the equilibrium value $\mathbf{p}_0^{\mathrm{sp}}(T)$ can be determined by minimizing the free energy $F(T, p_x, p_y, p_z)$ of the system. The initial assumption of a ferroic transition ensures that F can be determined at T_c.

In the vicinity of the transition, where $|\mathbf{p}_0^{\mathrm{sp}}|$ and $|T - T_c|$ are small, a Taylor expansion of F can be performed:

$$F(T, p_x, p_y, p_z) = F_0(T) + \sum_i \alpha_i(T) p_i + \sum_{ij} \beta_{ij}(T) p_i p_j + \dots \,. \tag{2.2}$$

The important point is that $F(T, p_x, p_y, p_z)$ is invariant under $\hat{G}_0$, because the free energy depends on the internal state of the system only and not on the absolute orientation. The same is valid for the scalar functions $\alpha_i(T)$ and $\beta_{ij}(T)$.

Considering the transformations of equation (2.2), we find that none of the linear terms is invariant under $\hat{G}_0$: The three reflections $\overline{2}_i$ ($i = x, y, z$) belonging to $4/mmm$ reverse the corresponding component p_i while leaving the others invariant, leading to a change of sign in the related linear term. Hence all $\alpha_i(T)$ are zero. In the same way we find that all cross-terms $p_i p_j$ ($i \neq j$) have to vanish.

Application of 4_z furthermore transforms p_x^2 to p_y^2, so that the associated coefficients $\beta_{ii} =: \beta_i$ have to be equivalent and we can write for the second-order approximation of the free energy F:

$$F(T, p_x, p_y, p_z) = F_0(T) + \frac{\beta_1(T)}{2}(p_x^2 + p_y^2) + \frac{\beta_2(T)}{2}p_z^2 \,. \qquad (2.3)$$

In the next step we address the specific form of the continuous functions $\beta_i(T)$: First, all $\beta_i(T)$ have to be positive above $T \geq T_c$. Otherwise equation (2.3) would not reach its minimum for $\mathbf{p}_0^{\text{sp}}(T) = 0$, contrary to the initial assumption of a paraelectric state.

To ensure that the minimum of F leads to a ferroelectric state ($\mathbf{p}_0^{\text{sp}}(T) \neq 0$) below T_c, one of the functions $\beta_i(T)$ has to become negative across the transition. In the case that $\beta_2(T)$ becomes zero at T_c, only the component p_z is affected by the transition, while the set (p_x, p_y) remains zero across the transition and therefore can be neglected in its description. The remaining component p_z, whose equilibrium value is modified by the transition, is the *order parameter* of the transition. But order parameters can also have more than one component. If $\beta_1(T)$ vanishes at T_c instead of $\beta_2(T)$, the relevant order parameter is given by the set (p_x, p_y), constituting a two-dimensional order parameter (see also figure 2.1(b)).

2.1.3 First- and second-order transitions

It was just discussed that the sign of the order-parameter-related function $\beta_i(T)$ changes sign at T_c. Therefore, the first term of a Taylor expansion of $\beta_i(T)$ as function of $(T - T_c)$ is:

$$\beta_i(T) = a(T - T_c) \,, \qquad (2.4)$$

with $a > 0$. The condition $\beta_i(T < T_c) < 0$ ensures that $\mathbf{p}_0^{\text{sp}} = 0$ is not a minimum of F below T_c.

However, it turns out that equation (2.3) is insufficient to describe a ferroelectric phase and even leads to unphysical results: In the ordered phase the free energy becomes infinitely negative for $|\mathbf{p}_0^{\text{sp}}| \to \infty$. To overcome this problem and regain a minimum of F at finite order-parameter values, higher-order terms have to be included.

In the case of an order parameter p_z the fourth order expansion of the free energy is

$$F(T, p_z) = F_0 + \frac{a}{2}(T - T_c)p_z^2 + \gamma_1 p_z^4 \,. \qquad (2.5)$$

As we will see in the following, the sign of the constant γ_1 plays a fundamental role for the type of the phase transition that occurs at T_c.

Second-order phase transitions: For a positive constant $\gamma_1 > 0$ equation (2.5) has a minimum at finite p_z for $T < T_c$. As depicted in figure 2.2(a), two minima of F occur at

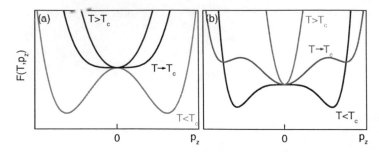

Figure 2.2: Temperature-dependent evolution of the free energy in the case of a second- and a first-order transition. (a) Form of the free energy F given by equation (2.5). Above T_c the free energy reaches its minimum for $p_z = 0$ (paraelectric). As the temperature is lowered ($T \to T_c$), the shape of the function changes until two minima ($p_z \neq 0$) arise for $T < T_c$ in a continuous way, constituting a second-order phase transition. (b) In the case of a first-order phase transition, local minima arise ($p_z \neq 0$) as T approaches T_c. They become energetically favorable at $T \leq T_c$ so that a discontinuous (jump-like) change of p_z occurs at the transition from zero to a finite value ($p_z \neq 0$).

$\pm p_z^{\mathrm{sp}}(T)$ with

$$p_z^{\mathrm{sp}}(T) = \sqrt{\frac{a(T_c - T)}{\gamma_1}} \, . \tag{2.6}$$

Obviously, $p_z^{\mathrm{sp}}(T)$ vanishes continuously for increasing temperature as T_c is approached. Hence the phase transition can be classified as a continuous or *second-order* phase transition.

The result is not restricted to ferroelectric phase transitions considered here but it can be expanded to any kind of order parameter. In conclusion, the general scaling behavior of order-parameter components η_i below a second-order phase transition is given by

$$\eta_i \propto (T_c - T)^{\frac{1}{2}} \, . \tag{2.7}$$

First-order phase transitions: In contrast, a negative constant $\gamma_1 < 0$ again requires higher-order terms to ensure a minimum of the free energy at finite values of p_z for $T < T_c$, so that F becomes

$$F(T, p_z) = F_0 + \frac{a}{2}(T - T_c)p_z^2 + \gamma_1 p_z^4 + \gamma_2 p_z^6 \, , \tag{2.8}$$

with $\gamma_2 > 0$. However, this modification of F qualitatively changes the character of the phase transition as demonstrated in figure 2.2(b).

The free energy F still reaches its minimum for $p_z = 0$ above T_c. But as the temperature is lowered, additional local minima show up ($p_z \neq 0$) which become energetically favorable compared to $p_z = 0$ for $T < T_c$. Consequently, a discontinuous (*first-order*) transition occurs from $p_z^{\mathrm{sp}} = 0$ to a finite value of $p_z^{\mathrm{sp}} \neq 0$.

2.1.4   Secondary order parameters

So far we considered transitions that are induced by a single order parameter. However, in general additional variables can arise, due to a coupling to the so-called *primary* order parameter that drives the transition. For example electrostrictive strain can accompany the ferroelectric transition discussed before. In this case the electrostrictive strain would be described by a *secondary* order parameter, expressing that the strain does not drive the transition. It is merely a consequence of the ferroelectric ordering, exclusively arising due to a coupling to the primary ferroic order parameter, i.e., the spontaneous polarization.

To include secondary order parameters in our formalism, we now allow for distance changes between planes of negative ions along the z axis of our tetragonal system. The change corresponds to a deformation of the unit cell, that can be represented by a component e' of the strain tensor $\hat{e}$. As before, the equilibrium value of e' can be deduced by minimizing $F(T, p_z, e')$ with respect to e' in addition to the variables already considered. Being a component of the symmetric second-rank strain tensor, e' is invariant under all symmetry operations of $\hat{G}_0$, so that any power of e' is allowed to contribute to F. The part of F depending only on e' is:

$$F(e') = \lambda e' + \frac{Ce'^2}{2} + \frac{De'^3}{3} + \dots , \qquad (2.9)$$

with coefficients λ, C, and D [46]. Apparently, the multiplication of equation (2.9) with an invariant polynomial of p_z is again invariant. Thus, such mixed terms can also contribute to the free energy in equation (2.3) and the lowest order coupling term between e' and the order-parameter p_z is given by $\delta e' p_z^2$, where the coupling constant δ accounts for the strength of the coupling.

However, to analyze the scaling behavior of the secondary order parameter e' further calculations are necessary. The reason is that in this special case the coefficient λ in equation (2.9) is connected to the thermal expansion which is generally not small. As a consequence, e' cannot assumed to be small and higher-order contributions to the free energy cannot be neglected[1].

This problem can be solved by setting $e' = e_0 + e$ with e_0 being the nonzero value of e' at the transition temperature. The introduced e_0 is defined by

$$\lambda + Ce_0 + De_0^2 + \dots = 0 \qquad (2.10)$$

and can be treated as a kind of background caused by thermal expansion which is not affected the by the transition itself. Both λ and e_0 vary smoothly across T_c and can be integrated into F_0. A short calculation reveals the resulting structure of the transition-relevant parameter e:

$$
\begin{aligned}
\frac{\partial F}{\partial e'} &= \lambda + Ce' + De'^2 + \delta p_z^2 \\
&= \lambda + C(e_0 + e) + D(e_0 + e)^2 + \delta p_z^2 \\
&= \underbrace{\lambda + Ce_0 + De_0^2}_{=0} + e\underbrace{(C + 2De_0)}_{=\text{const}} + \underbrace{De^2}_{\ll 1} + \delta p_z^2 \overset{!}{=} 0 .
\end{aligned}
\qquad (2.11)
$$

Contrary to e', the value e is small so that the second-order term De^2 can be neglected. This leads to the expression

$$e = -\frac{\delta}{C + 2De_0} \cdot p_z^2 , \qquad (2.12$$

---

[1]By differentiation of equation (2.9) with respect to e' one can see that the magnitude of λ is correlated to the magnitude of e'.

G_0	4_z	$\overline{2}_x$	$\overline{1}$
p_x	p_y	$-p_x$	$-p_x$
p_y	$-p_x$	p_y	$-p_y$
p_z	p_z	p_z	$-p_z$

Table 2.1: Action of the generators of $\hat{G}_0$ on the possible order-parameter components p_x, p_y, and p_z.

determining the secondary order parameter e of the phase transition. Equation (2.12) implies that the secondary order parameter only appears in the case of a nonzero coupling constant δ and reveals the general scaling behavior of secondary order parameters.
Since $p_z \propto (T_c - T)^{\frac{1}{2}}$ (see equation (2.6)), the secondary order parameter exhibits a linear dependence on $(T_c - T)$:

$$e \propto (T_c - T) . \tag{2.13}$$

Hence, primary and secondary order parameters can usually be identified by their scaling behavior, provided that only one primary order parameter couples to the secondary order parameter (see also section 5.1.3).
Any kind of ferroic order that is associated to a secondary order parameter is called an *improper* ferroic quality, while the term *proper* is used to indicate a relation to a primary order parameter. Accordingly, we distinguish between improper and proper ferroic transitions [46,48].
However, a coupling between primary and secondary order parameters does not affect the principle structure of the equation describing the free energy (equation (2.5)). Thus, none of the derived conclusions is affected by the presence of nonzero secondary order parameters. Stable phases and types of transitions (first- or second-order) are the same as in the absence of a coupling. Only phase boundaries will slightly shift in temperature, because some coefficients in the expansion of F change.
The onset of a secondary order parameter always conserves the symmetry $\hat{G} \subset \hat{G}_0$ imposed by the primary order parameter and we can summarize:
The symmetry characteristics of a transition are *exclusively* determined by the primary order parameter, whereas secondary order parameters are merely a "byproduct".

2.1.5 Representation theory of order parameters

After discussing different types of phase transitions and introducing the concept of order parameters, we now address their description in terms of group theory. The group-theoretical treatment provides general symmetry-based information about the total number of possible order parameters or the quantity of order-parameter components. Important definitions are introduced on the basis of the previous example, i.e., a ferroelectric transition in a crystal with point symmetry $\hat{G}_0 = 4/mmm$.
In this case, either p_z or (p_x, p_y) can be the relevant primary order parameter regarding the ferroelectric phase transition at T_c (see equation (2.3)). Hence, the order parameter driving the transition can have one or two components.
The reason is that some symmetry operations of the point group $\hat{G}_0$ exchange p_x and p_y, but no symmetry operation exchanges p_x or p_y with p_z. This is shown in detail in table 2.1 for the generators 4_z, $\overline{2}_x$, and $\overline{1}$. Since all symmetry operations of $\hat{G}_0$ can be represented by a chosen set of generators, it is sufficient to regard their action on (p_x, p_y, p_z).

G_0	4_z	$\bar{2}_x$	$\bar{1}$
(p_x, p_y, p_z)	$\begin{pmatrix} 0 & -1 & 0 \\ 1 & 0 & 0 \\ 0 & 0 & 1 \end{pmatrix}$	$\begin{pmatrix} -1 & 0 & 0 \\ 0 & 1 & 0 \\ 0 & 0 & 1 \end{pmatrix}$	$\begin{pmatrix} -1 & 0 & 0 \\ 0 & -1 & 0 \\ 0 & 0 & -1 \end{pmatrix}$
(p_x, p_y)	$\begin{pmatrix} 0 & -1 \\ 1 & 0 \end{pmatrix}$	$\begin{pmatrix} -1 & 0 \\ 0 & 1 \end{pmatrix}$	$\begin{pmatrix} -1 & 0 \\ 0 & -1 \end{pmatrix}$
p_z	1	1	−1

Table 2.2: Reducible representations of the generators of $\hat{G}_0 = 4/mmm$ and irreducible representations of the vector spaces, carried by the order parameters belonging to $\hat{G}_0$ (see text).

Table 2.1 indicates, that (p_x, p_y) can be considered as a vector of a two-dimensional vector space $\mathcal{V}_2$, being invariant under $\hat{G}_0$: Regardless which symmetry operation of $\hat{G}_0$ is applied to (p_x, p_y), the resulting vector is again an element of $\mathcal{V}_2$. The same is valid regarding (p_z) as a vector of a one-dimensional vector space $\mathcal{V}_1$.

In general this is expressed by saying that the n components of an order parameter span a n-dimensional vector space, which is invariant with respect to the point group $\hat{G}_0$. The existence of these invariant vector spaces has important consequences for the group-theoretical treatment.

First of all, the action of a symmetry operation on (p_x, p_y, p_z) can be expressed or *represented* by a (3×3)-matrix as shown for the generators of $4/mmm$ in table 2.2. These matrices are called a *reducible representation* of $\hat{G}_0$.

However, as mentioned above, smaller invariant subspaces $\mathcal{V}_2$ and $\mathcal{V}_1$ exist. Whenever this is the case, the reducible represenation can be transformed to a block diagonal matrix by choice of an appropriate basis [49]. In our example the "right" basis is given by a simple Cartesian coordinate system, determined by the crystal axes x, y, and z. Therefore the three-dimensional matrices (the reducible representation) already have block form (indicated by the straight lines in table 2.2). The dimension of the different blocks is detemined by the dimension of the existing smaller invariant vector spaces, here denoted by $\mathcal{V}_2$ and $\mathcal{V}_1$.

The two-dimensional and one-dimensional matrices forming the blocks of the (3×3)-matrices are called *irreducible representations* of $\hat{G}_0$, being related to (p_x, p_y) and (p_z), respectively (see table 2.2). The important point is, that this fact can be generalized, meaning that an order parameter of dimension n is always related to an irreducible representation of the same dimension.

Conversely, the knowledge of all irreducible representations related to (p_x, p_y, p_z) under $\hat{G}_0$ immediately allows one to make a statement about possible order parameters and the number of their components. The same is in principle valid for incommensurate structures, although certain modifications are needed.

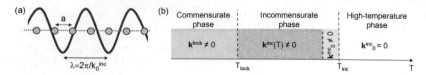

Figure 2.3: General scheme of incommensurate phase transitions. The ordering right below T_{inc} is determined by a temperature-independent incommensurate wave vector $\mathbf{k}_0$, followed by a region described by a T-dependent $\mathbf{k}^{inc}(T)$. At even lower temperatures a *lock-in* transition to a commensurate orderd phase with $\mathbf{k}^{lock}$ takes place.

2.2   Incommensurate phases

Strongly correlated electronic materials often show complex long-range magnetic ordering. Im many cases competing magnetic interactions lead to a so-called *incommensurate* modulation of the magnetic moments.

In general, electronic structures, displacements of atoms or magnetic modulations are said to be *incommensurate* when the ratio of their periodicity and the periodicity of the crystal lattice is no rational number.

This is illustrated in figure 2.3(a), schematically showing an one-dimensional chain of atoms and a sinusoidal modulation, which is incommensurate with respect to the arrangement of atoms.

In three dimensions the wave vector (also propagation vector) $\mathbf{k}^{inc}$ describing such an incommensurate modulation has at least one irrational component. Because of the incommensurability, the resulting structure lacks translation symmetry along those crystallographic axes that are related to an irrational component of $\mathbf{k}^{inc}$ [50].

The important difference between modulated structures (commensurate or incommensurate) and the simple example discussed in the previous section is, that the order parameter η of a periodically modulated phase is related to a nonzero wave vector. Whenever this is the case, also the symmetry properties of the wave vector have to be taken into account, because the associated modulation will lower the symmetry of the system[2].

To include the characteristic features of periodically modulated phases, especially incommensurate phases, we have to modify the formalism introduced in the previous sections. In the present case, the space group G_0 of a crystal has to be considered instead of its point group $\hat{G}_0$. The modification ensures that also incommensurate structures can be treated in terms of the Landau theory, while the aforementioned general principles regarding order parameters and the evaluation of stable ordered phases remain valid.

2.2.1   Phenomenology of incommensurate transitions

Before we come to the details regarding the group theoretical treatment of incommensurate structures, the generalized phase diagram of an incommensurate system is discussed. It is schematically depicted in figure 2.3(b).

The phase diagram consists of three phases, separated by two transitions, with the upper one at T_{inc} being the incommensurate transition [46]. Right below T_{inc} the physical pro-

---

[2]The ferroelectric order discussed before can be described by a wave vector $\mathbf{k} = 0$, indicating an "infinite" period of the modulation. For $\mathbf{k} = 0$ it is sufficient to analyze the point-group symmetry of the ordered system.

perties can be represented by a standard primary order parameter. This order parameter is related to an incommensurate wave vector $\mathbf{k}_0^{\text{inc}}$, having at least one irrational component q_i^0. As long as this wave vector remains constant, the situation is quite similar to ordinary ferroic transitions discussed before and the order parameter is related to a single irreducible representation.

Further away from T_{inc} the order parameter may not remain associated to the same irreducible representation as right below T_{inc}. The reason is that the components of an incommensurate wave vector are often temperature dependent ($q_i^{\text{inc}}(T)$). Changes in $\mathbf{k}^{\text{inc}}(T)$ possibly affect the symmetry of the system, leading to different associated irreducible representations at different temperatures. In general, only right below T_{inc} the assumption of a temperature-independent $\mathbf{k}_0^{\text{inc}}$ is a good approximation.

The incommensurate phase with $\mathbf{k}^{\text{inc}}(T)$ is usually stable down to a second transition at T_{lock}, called the *lock-in* temperature. Below T_{lock} the system is in its commensurate ground state with temperature-independent wave vector values q_i^{lock}. Hence, in the whole temperature range $T < T_{\text{lock}}$ the order parameter is again associated to one and the same irreducible representation.

The values of q_i^{lock} are close to the values q_i^0, but they are rational numbers. The rational numbers simply indicate that the primitive translations of the commensurate phase are multiples of the primitive translations in the high-temperature phase. Therefore the low-temperature commensurate phase typically has a larger unit cell and a lower symmetry than the crystal structure.

The basic idea of the extended Landau theory for incommensurate phases is to consider two types of contributions F_{inc} and F_{lock} to the free energy. The first contribution expresses the tendency towards an incommensurate ordering with wave vector $\mathbf{k}_0^{\text{inc}}$ right below T_{inc}, whereas the second one prefers commensurate ordering, associated to $\mathbf{k}^{\text{lock}}$.

To ensure that the behavior of the system right below T_{inc} (where the order parameter η is small) is dominated by F_{inc}, the terms in F_{lock} are of higher powers in the order parameter η compared to F_{inc}. Thus, the free energy F could be for example of the form

$$F(T) = F_0(T) + \underbrace{\eta^2 g_{\text{inc}}}_{F_{\text{inc}}} + \underbrace{\eta^4 g_{\text{lock}}}_{F_{\text{lock}}} + \dots , \qquad (2.14)$$

with functions g_{inc} and g_{lock} describing an incommensurate and commensurate modulation of the ordering, respectively. As the temperature is lowered the order parameter η increases and the F_{lock} term ($\propto \eta^4$) becomes more important, finally inducing the transition at T_{lock}. However, it was already mentioned that modulated structures require a symmetry analysis in terms of space groups. Unfortunately, the entire incommensurate phase is, strictly speaking, not a crystalline state. Although the vector $\mathbf{k}_0^{\text{inc}}$ can still be associated to one primary order parameter and reflects the existence of a perfectly ordered state, it has irrational components q_i^0. The incommensurate modulation violates the three-dimensional translation symmetry of the system and therefore prohibits a description of the incommensurate phase by a three-dimensional space group.

This problem can be solved for instance by analyzing the incommensurate structure in terms of higher-dimensional group theory [51–53].

Another approach is to identify the incommensurate phase by a part of the symmetry properties of the primary order parameter η remaining in the incommensurate phase [46]. This means in particular that the high-temperature phase ($T > T_{\text{inc}}$) and the ground state ($T < T_{\text{lock}}$) are described by *space groups*, whereas the intermediary incommensurate phase can only be specified by its *point-group* symmetry.

Here we focus on the latter approach which is outlined in the next section.

2.2.2 Space-group representation

The analysis of order parameters of periodically modulated phases is based on the space-group symmetry of the investigated crystal. The representation of space groups is more complex compared to the point-group representation, because point *and* translation symmetries have to be considered. In the following, we discuss the standard procedure for the construction of spatial group representations. The objective is to find the irreducible representations of a crystal exhibiting incommensurate ordering.

On the one hand these irreducible representations are needed to deduce the form of the free energy, i.e., to evaluate the order-parameter invariants. On the other hand, the irreducible representations directly provide information about the dimension of incommensurate order parameters.

Translation group T

The translation group T is a subgroup of the space group G_0 and its elements $T_t \in T$ can be associated to the lattice points

$$\mathbf{t} = m_1\mathbf{a_1} + m_2\mathbf{a_2} + m_3\mathbf{a_3} \quad (m_i = 0,\ \pm 1,\ \pm 2,\ \dots) \tag{2.15}$$

of a Bravais lattice.

Regarding a generic function $f(\mathbf{r})$, application of T_t simply leads to a shift in the argument, defined by

$$T_t f(\mathbf{r}) = f(\mathbf{r} - \mathbf{t}) . \tag{2.16}$$

Apparently, a general translation operation decomposes in three commuting one-dimensional translations in the directions $\mathbf{a}_i$ denoted by [54]

$$T_t = T_{t_1} T_{t_2} T_{t_3} , \tag{2.17}$$

with $\mathbf{t}_i = m_i \mathbf{a}_i$. Each of three one-dimensional translations can be represented by a phase factor or character

$$e^{ik_i m_i a_i} .$$

These phase factors are defined by the components of $\mathbf{k}$, being a vector of the first Brillouin zone. Accordingly, a general three-dimensional translation T_t is associated to the phase factor

$$e^{ik_1 m_1 a_1} e^{ik_2 m_2 a_2} e^{ik_3 m_3 a_3} = e^{i\mathbf{k}\mathbf{t}} ,$$

which can be used to construct the irreducible representations $\Gamma^{(\mathbf{k})}$ of T [55]. Each single wave vector $\mathbf{k}_i$ of the first Brillouin zone gives rise to an irreducible representation of the translation group denoted by

$$
\begin{array}{c|c|c|c|c|c}
T_t & [100] & [010] & \cdots & \mathbf{t} & \cdots \\
\hline
\Gamma^{(\mathbf{k}_1)} & e^{i\mathbf{k}_1 \mathbf{a}_1} & e^{i\mathbf{k}_1 \mathbf{a}_2} & \cdots & e^{i\mathbf{k}_1 \mathbf{t}} & \cdots \\
\hline
\Gamma^{(\mathbf{k}_2)} & e^{i\mathbf{k}_2 \mathbf{a}_1} & e^{i\mathbf{k}_2 \mathbf{a}_2} & \cdots & e^{i\mathbf{k}_2 \mathbf{t}} & \cdots \\
\vdots & & & & &
\end{array}
\tag{2.18}
$$

Since the number of $\mathbf{k}$–points in the first Brillouin zone is equivalent to the number of unit cells, there are $\approx 10^{23}$ such irreducible representations $\Gamma^{(\mathbf{k}_i)}$ of T for a finite crystal and an infinite number of $\Gamma^{(\mathbf{k}_i)}$ for a Bravais lattice [56].

However, the important point is that one irreducible representation can always unambiguously be labelled by the wave vector $\mathbf{k}$. Reversly, a given wave vector $\mathbf{k}$ identifies one specific irreducible representation of T.

Note that the restriction to the first Brillouin zone is important. Otherwise, the irreducible representations are not well-defined, because two wave vectors $\mathbf{k}$ and $\mathbf{k}'$ are equivalent whenever their difference is equal to a vector $\mathbf{K}$ of the reciprocal lattice [57]

$$\mathbf{k}' - \mathbf{k} = \mathbf{K} , \tag{2.19}$$

leading to

$$\Gamma^{(\mathbf{k}')} = e^{-i\mathbf{k}'\mathbf{t}} = e^{-i\mathbf{k}\mathbf{t}} \underbrace{e^{-i\mathbf{K}\mathbf{t}}}_{=1} = \Gamma^{(\mathbf{k})} . \tag{2.20}$$

For incommensurate phases the translation symmetry of finite crystals is violated. The associated wave vector $\mathbf{k}^{\text{inc}}$ still indicates a perfectly ordered phase, but the translations determined by $\mathbf{k}^{\text{inc}}$ cannot be represented by primitive translations of the lattice (see table 2.18).

Thus, contrary to the high-temperature phase and the commensurate ground state of the schematic phase diagram in figure 2.3(b), no space group $G \subset G_0$ exists that describes the intermediary incommensurate phase.

Little group of the wave vector

We now consider the transformation of a general wave vector $\mathbf{k}$ under pure point operations R of the space group G_0.[3]

Since $\mathbf{k}$ is a vector of the reciprocal lattice, it is worth mentioning that the reciprocal lattice is invariant under the same point-group operations as the real-space structure [58].

The set of elements R that transforms $\mathbf{k}$ into itself or into an equivalent wave vector in the sense of equation (2.19) forms a subgroup $G(\mathbf{k})$ of G_0. This group $G(\mathbf{k})$ is called the *little group* of $\mathbf{k}$. It is a point group and its irreducible representations $\tau_j(\mathbf{k})$ are listet for instance in reference [59]. They are called the *small* representations.

Apparently, different wave vectors belong to different little groups. Thus, $\mathbf{k}$ can be identified by one single irreducible representation $\tau_j(\mathbf{k})$ of $G(\mathbf{k})$. We have already seen in section 2.1.5 that the dimension of $\tau_j(\mathbf{k})$, denoted by $\dim(\tau_j(\mathbf{k}))$, is equivalent to the dimension of an invariant subspace $\mathcal{V}_j$ regarding the reducible matrix representation of the group $G(\mathbf{k})$.

Figure 2.4(a) schematically shows the point operations belonging to the little group of $\mathbf{k}$, considering a point Σ lying on a mirror plane of the first Brillouin zone of a square lattice. Here, only the two operations 1 and $\overline{2}_z$ belong to the little group $G(\mathbf{k})$.

In figure 2.4(b) the so-called *star of* $\mathbf{k}$ is depicted. The star of $\mathbf{k}$ consists of all nonequivalent wave vectors generated by point operations of G_0. Regarding the point Σ of the square lattice, 1, 2_z, and $\pm 4_z$ are operators that lead to nonequivalent wave vectors. Each wave vector constitutes an *arm of the star*.

---

[3]The presented analysis applies to so-called symmorphic space groups, where all symmetry operations apart from the translations leave one lattice point fixed. Non-symmorphic space groups also include combinations of non-primitive translation with point operations as screw-rotations. Hence, no point of the lattice is fixed in the latter case. In the case of non-symmorphic space groups $\tau_j(\mathbf{k})$ is no irreducible representation in the usual sense, because it also depends on the translations in the space group [45]. Nevertheless, the basic results deduced in this section are valid in both cases.

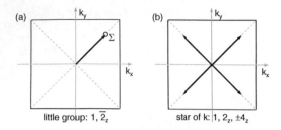

Figure 2.4: (a) Point-symmetry operations of the little group of the wave vector **k** that is related to a point Σ which lies on a mirror plane of the square lattice. (b) The star of **k** associated to Σ consists of four elements called the *arms of the star*.

Irreducible representations of a space group

The definition of the little group and the star of **k** allows to decompose a general point operation R of the space group G_0 into two operations

$$R = G \circ H ,$$

with $G \in G(\mathbf{k})$ and H denoting an operator of the star of **k**. The small representations $\tau_j(\mathbf{k})$ are known [59] and each arm of the star is associated to a different small representation, because nonequivalent **k** belong to different little groups.
Therefore, G_0 contains ℓ invariant subspaces of dimension $\dim(\tau_j(\mathbf{k}))$, with ℓ denoting the number of arms in the star.
Hence, we can conclude that the dimension of the irreducible representations $\Gamma_j^{(\mathbf{k})}$ of $R \in G_0$ is given by

$$\dim(\Gamma_j^{(\mathbf{k})}) = \ell \cdot \dim(\tau_j(\mathbf{k})) . \tag{2.21}$$

Apparently, an irreducible representation $\Gamma_j^{(\mathbf{k})}$ of a space group G_0 is specified by two elements:

1. A star of **k** (different stars define different irreducible representations of G_0).

2. One small representation $\tau_j(\mathbf{k})$ of the little group (different $\tau_j(\mathbf{k})$ define different nonequivalent irreducible representations of G_0).

This means for instance that if the little group $G(\mathbf{k})$ has two one-dimensional small representations ($\tau_1(\mathbf{k})$ and $\tau_2(\mathbf{k})$) and if the number of arms in the star of **k** is $\ell = 4$, there will be two four-dimensional irreducible representations of G_0, denoted by $\Gamma_1^{(\mathbf{k})}$ and $\Gamma_2^{(\mathbf{k})}$. Both elements, i.e. the star of **k** and the small representations $\tau_{1,2}(\mathbf{k})$, can easily be calculated by using the *Bilbao Crystallographic Server* [60].
Note that the dimension of the irreducible representations $\Gamma_{1,2}^{(\mathbf{k})}$ is equivalent to the dimension of the associated order parameters $\eta_{1,2}$, irrespective of **k** being a commensurate or an incommensurate wave vactor. The difference between commensurate and incommensurate wave vectors simply manifests in the associated little groups and different stars.
One special feature of incommensurate structures is that the star always consists of wave vector pairs. This is because the Brillouin zone is a centrosymmetric body. Hence, with **k** also $-\mathbf{k}$ is a vector of the Brillouin zone. Since incommensurate wave vectors **k** and

$-\mathbf{k}$ are not equivalent in the sense of equation (2.19), both vectors belong to the star of $\mathbf{k}$ ($\pm\mathbf{k} \Rightarrow \ell = 2, 4, 6, ...$).

Thus, according to equation (2.21), the simplest possible incommensurate order parameter η is two-dimensional.

Conclusion

It was shown that the irreducible representations $\Gamma_j^{(\mathbf{k})}$ of a space group can be determined on the basis of the wave vector $\mathbf{k}$. For incommensurate phases described by $\mathbf{k}^{\text{inc}}$ the translation symmetry of the high-temperature phase with space group G_0 is violated.

Thus, incommensurate phases cannot be described by a space group. However, some of the point operations belonging to G_0 remain symmetry operations for incommensurate phases, so that incommensurate phases can at least be specified by denoting their point-group symmetry.

2.2.3   Free energy and incommensurate order parameters

The discussion of the space-group representation revealed that the simplest order parameter associated to an incommensurate structure is two-dimensional, being denoted by (η_1, η_2). Especially in the context of periodic structures it is helpful to express the order parameter by complex components. An equivalent complex order parameter to (η_1, η_2) can be defined by noting [61]

$$\eta = \eta_1 + i\eta_2 \tag{2.22}$$
$$\eta^* = \eta_1 - i\eta_2 . \tag{2.23}$$

By introducing the polar coordinates (ρ, θ), the order parameter becomes

$$\begin{pmatrix} \eta_1 \\ \eta_2 \end{pmatrix} \Leftrightarrow \begin{pmatrix} \eta \\ \eta^* \end{pmatrix} = \begin{pmatrix} \rho e^{i\theta} \\ \rho e^{-i\theta} \end{pmatrix} . \tag{2.24}$$

Thus, in the case of an incommensurate spin structure the complex amplitudes of the magnetic waves can be taken as the complex order-parameter components.

Once the irreducible representations of an incommensurate system are known, the free energy $F(T, \rho, \theta)$ is derived by finding the most general polynomials in the order-parameter components η_1 and η_2 which are invariant with respect to the irreducible representations $\Gamma_j^{(\mathbf{k})}$ (η_1 and η_2 are now assumed to be complex).

Just as discussed in section 2.1.2, minimization of the free-energy expansion $F(T, \rho, \theta)$ with respect to ρ and θ reveals the symmetry and stability conditions of possible ordered phases. However, with respect to a later discussion of symmetry-breaking incommensurate order parameters, it is important to recognize that the free energy is partially independent of the phase θ in η_1 and η_2, given by equation (2.24). The independence implies that the phase can be redefined without changing the free energy of the system. Taking for example a sinusoidal spin density wave (see figure 3.5), this means that the phase of the spin arrangement can always be redefined, so that the magnetic and the crystallographic center of inversion coincide. Thus, a sinusoidal spin density wave conserves spatial inversion symmetry.

This can be shown mathematically, regarding a general term in the free energy, which is a sum of expressions like

$$\eta_1^a \eta_2^b = \rho^{a+b} e^{i\theta(a-b)} , \tag{2.25}$$

where η_1 and η_2 are the complex order-parameter components and a and b denote their power in the free energy expansion. To investigate the meaning of the order-parameter phase θ, we first consider the action of a simple translation matrix $[\mathbf{a}_j]$ on $\eta_{1,2}$, using the relation $\mathbf{a}_i^*\mathbf{a}_j = 2\pi\delta_{ij}$:

$$[\mathbf{a}_j]\eta_{1,2} = e^{\mp\mathbf{k}_0\mathbf{a}_j}\eta_{1,2} = e^{\mp i\sum_i q_i^0 \mathbf{a}_i^*\mathbf{a}_j}\eta_{1,2} = e^{\mp 2\pi i q_j^0}\eta_{1,2} \ . \tag{2.26}$$

Applying the same translation to the general term of the free energy (equation (2.25)) leads to

$$([\mathbf{a}_j]\eta_1)^a ([\mathbf{a}_j]\eta_2)^b = \rho^{a+b}e^{i\theta(a-b)}e^{-2\pi i q_j^0(a-b)} = \eta_1^a\eta_2^b e^{-2\pi i q_j^0(a-b)} \ . \tag{2.27}$$

One can immediately see, that the necessary invariance of this term imposes the condition

$$e^{-2\pi i q_j^0(a-b)} = 1 \ . \tag{2.28}$$

Since q_j^0 is an irrational number for incommensurate ordering, this can only be achieved by $a = b$, meaning that any term of the free energy described by equation (2.25) has to be independent of the angular variable θ.

Note that the phase-angle independence is only valid as long as only one nonzero order parameter is present. As soon as a second order parameter arises, the relative phase between the two order parameters becomes relevant for the system.

Chapter 3

Multiferroic materials

This chapter covers a brief review about materials with coexisting magnetic and ferroelectric order, called multiferroics. After introducing a helpful classification of these systems with respect to their inherent order parameters, we focus on those multiferroics where the spontaneous electric polarization forms as a direct consequence of long-range magnetic ordering, namely magnetically induced ferroelectrics.

Magnetically-induced ferroelectrics are of special interest, because they exhibit a particularly strong coupling between the magnetic and the ferroelectric properties. Since no textbooks are available dealing explicitly with the physics of this class of multiferroic materials, possible mechanisms and proposed models are discussed in detail.

In the last section of this chapter, the concept of domains in ferroic materials is addressed, focussing on possible domain structures in magnetically induced ferroelectrics.

3.1 Multiferroics

Multiferroics were formally defined by H. Schmid in 1994 [62]. He proposed the following rigorous definition for single-phase crystals:

> *Crystals can be defined as multiferroic when two or more*
> *of the primary ferroic properties are united in the same phase.*

The primary ferroic properties addressed here are ferroelectricity, ferromagnetism, ferroelasticity and ferrotoroidicity, as sketched in figure 3.1. The term "ferroic" simply means that the relevant order parameter can be switched under external influence (electric field, magnetic field, mechanical stress) into at least two positions [64,65].

In conventional ferroelectrics, microscopic electric dipoles align and form a macroscopic polarization that can be reversed by applying an external electric field. In ferromagnets it is the net magnetization, in ferroelastics the spontaneous strain, and in ferrotoroidics the alignment of magnetic vortices that can be switched.

However, nowadays the definition is usually expanded towards antiferromagnetic or ferrimagnetic order and also composite materials or heterostructures are called multiferroic whenever its constituents unify two different ferroic properties. One example are magnetic $CoFe_2O_4$ pillars in a ferroelectric $BaTiO_3$ matrix [66].

In all cases, symmetry plays a fundamental role. Figure 3.1 summarizes the behavior of the four primary ferroic properties under space $(\mathbf{r} \rightarrow -\mathbf{r})$ and time $(t \rightarrow -t)$ reversal [63]. It is shown that a ferroelectric polarization is invariant under the time-inversion operation,

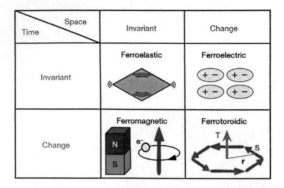

Figure 3.1: All forms of primary ferroic order under the parity operations of space and time. While ferromagnetic order violates time inversion symmetry but remains invariant under the space inversion operation, the situation is exactly the opposite for a ferroelectric system [63].

but breaks the spatial inversion symmetry. In contrast, ferromagnetic order violates time-inversion symmetry and is conserved under inversion of space. This can be understood considering the magnetic moments in a simple classical picture as depicted in figure 3.1, i.e., assuming that a magnetic moment is induced by a circular current. For $t \rightarrow -t$ the direction of the current changes and thus leads to a magnetic moment of opposite sign. Therefore, neither the reversal of time nor the inversion of space can be a symmetry operation for a system with coexisting ferromagnetic and ferroelectric ordering.

The close link between symmetry and (anti)ferroic ordering is the basis for the experimental investigations presented in this thesis and will be discussed in detail in section 4.

3.2   Magnetoelectric multiferroics

Multiferroics that exhibit a cross-coupling between their ferroelectric and their magnetic properties are of special interest. These systems are promising materials regarding the manipulation of magnetic properties by electric fields (and the reverse) — a highly desired property in microelectronics and spintronics [16].

They are often referred to as magnetoelectric multiferroics, although the term is somehow misleading. So-called magnetoelectric multiferroics do not mandatorily display the *linear* magnetoelectric effect, i.e., a polarization response $\mathbf{P}^{\mathrm{ind}}$ to an applied magnetic field $\mathbf{H}$, or conversely a spin response $\mathbf{M}^{\mathrm{ind}}$ to an applied electric field $\mathbf{E}$ according to [67–71]

$$\mathbf{P}^{\mathrm{ind}} = \hat{\alpha}\mathbf{H} \quad \text{or} \quad \mathbf{M}^{\mathrm{ind}} = \hat{\alpha}\mathbf{E} . \qquad (3.1)$$

Here, $\hat{\alpha}$ denotes the axial second rank tensor of the linear magnetoelectric effect (in Gaussian units). One example for a multiferroic material that is not a magnetoelectic is hexagonal $YMnO_3$ [72]. In turn, materials displaying the linear magnetoelectric effect do not have to be multiferroic. This is the case for Cr_2O_3 [73,74].

Keeping this in mind, we now briefly review the class of magnetoelectric multiferroics. First of all, two fundamentally different types of magnetoelectric multiferroics can be distinguished with respect to the inherent magnetic and electric order parameters. They

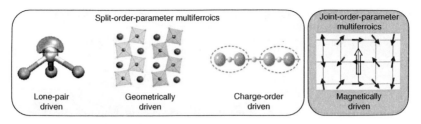

Figure 3.2: Magnetoelectric multiferroics: Four basic mechanisms are known to generate ferroelectricity in the class of magnetoelectric multiferroics. A fundamental difference regarding the correlation of involved order parameters suggests a further refined classification into split-order-parameter and joint-order-parameter multiferroics. [29, 75–77]

can either arise at different transition temperatures or in a coupled way at the same temperature, suggesting a classification as split-order-parameter multiferroics and joint-order-parameter multiferroics, respectively.

Both types of magnetoelectric multiferroics and possible mechanisms for the induced ferroelectricity are sketched in figure 3.2.

3.2.1 Split-order-parameter multiferroics

In *split-order-parameter* multiferroics the electric and the magnetic order parameters arise at different transition temperatures and due to different microscopic interactions. Although the critical temperatures of the magnetic and ferroelectric transitions can be well above room temperature, the coupling between the two properties is usually rather weak [34].

The microscopic origin of magnetism is basically the same in all split-order-parameter multiferroics. It originates from localized electrons in partially filled d or f shells of transition-metal ions or rare-earth ions [35]. The partial filling leads to local magnetic moments and exchange interactions that induce different types of magnetic ordering (ferro-, ferri- or antiferromagnetic).

Note that the d orbitals in the common ferroelectric materials as $BaTiO_3$ or $PbTiO_3$ are not occupied so that no magnetic moments exist. Consequently, no magnetic ordering occurs [78].

Regarding split-order-parameter systems, mainly three microscopically different mechanisms are known that lead to ferroelectric ordering:

Ferroelectricity due to lone pairs

The so-called lone-pair mechanism is based upon the violation of the local inversion symmetry by valence electrons.

Two well-known multiferroics of this type are $BiMnO_3$ and $BiFeO_3$, both belonging to the perovskite family. In either of the two compounds, ordering of the ferroelectric and the magnetic subsystem occur independent of each other, already reflected by the different transition temperatures T_{FE} and T_M, respectively ($BiFeO_3$: $T_{FE} = 1103$ K [79], $T_M = 643$ K [80]; $BiMnO_3$: $T_{FE} = 760$ K, $T_M = 105$ K [81]).

Their magnetic properties are related to the presence of magnetic transition-metal ions, i.e. Fe^{3+} (d^5) or Mn^{3+} (d^4), whereas the instability towards a ferroelectric ordering is driven by the Bi^{3+} ion on the A-site of their perovskite structure (see reference [82] for the denotation of the perovskite structure). Two of the Bi^{3+} valence electrons do not participate in chemical (sp)-hybridized states and create local dipoles. Below the ferroelectric transition temperature T_{FE} these local dipoles order and lead to a macroscopic net polarization [83].

Geometric ferroelectricity

In the case of geometrically-driven ferroelectricity, structural instabilities due to size effects or geometrical constraints are the origin of polar displacements.

Examples can be found in the series of hexagonal rare-earth manganites $RMnO_3$ with $R = $ Sc, Y, In, Dy – Lu[1]. Here the space-filling and ionic coordination in the high-temperature structure ($T_{FE} > 570$ – 990 K, depending on the particular R ion) is not optimal, but can be improved by a small distortion, leading to ferroelectric ordering [84].

The Mn^{3+} ions of hexagonal $RMnO_3$ are in a fivefold coordination, situated in the center of O_5 trigonal biprisms. In a two-step process, these rigid MnO_5 trigonal biprisms tilt at the ferroelectric transition to achieve close packing [85]. The tilting is accompanied by displacements of the R ions, leading to a loss of inversion symmetry and ferroelectric behavior, mostly caused by dipolar R – O pairs.

The ordering of the magnetic subsystem is completely decoupled from the structural ferroelectric phase transition at T_{FE}. Only below $T_M = 60$ to 130 K antiferromagnetic ordering of the Mn^{3+} moments is observed.

As a consequence, the coupling between the magnetic and electric properties is typically small. Again, $YMnO_3$ is an instructive example. Here, linear magnetoelectric interactions are forbidden by symmetry (all tensor components of $\hat{\alpha}$ are identically zero) and the coupling solely originates from an interaction between magnetic and electric domain walls [72].

Ferroelectricity due to charge ordering

In charge ordered insulators, ferroelectricity can appear as a consequence of non-centrosymmetric charge distributions [29].

Multiferroicity due to charge ordering includes a variety of mechanisms, being reviewed in reference [86]. Here, we only mention one well-established example to show the general character of a charge-ordered system. In the high-temperature paraelectric phase of $LuFe_2O_4$, the Fe ions have an average valency of $+2.5$. Below $T_{FE} = 330$ K charge ordering creates alternating layers with Fe^{2+}:Fe^{3+} ratios of 2:1 and 1:2, so that a spontaneous polarization between the two layers arises [87, 88]. Altogether, the charge-ordered polar bilayers lead to the observed macroscopic polarization in $LuFe_2O_4$.

The system becomes multiferroic as a long-range magnetically ordered phase is entered below $T_M \approx 240$ K. The Fe^{2+} ($S = 2$) and Fe^{3+} ($S = 5/2$) ions are in their high-spin configuration and order in a ferrimagnetic fashion [89, 90].

As common for split-order-parameter multiferroics, the magnetic and the electric subsystem in charge-ordered multiferroics order separately at significantly different temperatures and a generally weak cross-correlation between the two properties is observed in this type of multiferroics.

---

[1]Note that none of the hexagonal $RMnO_3$ systems displays the linear magnetoelectric effect.

3.2.2   Joint-order-parameter multiferroics

Maybe the most interesting magnetoelectric multiferroics are those, where complex magnetic long-range order simultaneously violates space-inversion and time-reversal symmetry, so that a spontaneous polarization is induced. They are referred to as *joint-order-parameter multiferroics*[2].

The denotation implies, that a magnetic (primary) and an electric (secondary) order parameter show up at one and the same phase transition. As the ferroelectricity appears only in certain magnetically ordered states, it is not surprising that the coupling between magnetic and electric properties is intrinsically strong in these materials. It is the inherent magnetic ordering that generates the polarization and thus provides a naturally pronounced magnetoelectric coupling.

This unconventional mechansim, i.e. a magnetically induced ferroelectric polarization, enables giant and spectacular effects, like the magnetic-field induced polarization flop in $TbMnO_3$ or the reversible magnetic-field control of an electric polarization in $TbMn_2O_5$ [8, 9]. It was mainly the observation of these two effects, that excited the broad current interest in multiferroic materials, although early examples as Cr_2BeO_4 [94] were known decades before.

Many other systems, belonging to the class of joint-order-parameter multiferroics, are known nowadays and the number is still increasing. Unfortunately, most of the systems are far from being understood, due to the complexity of their magnetism and the intricate interactions between the spins, charge and lattice.

Special highlights besides the already mentioned systems are the ferrimagnetic spinel oxide $CoCr_2O_4$ [13], the helimagnet $Ba_2Mg_2Fe_{12}O_{22}$ [15] or the spin-spiral system CuO [32]. While the spinel $CoCr_2O_4$ allows for an electric-field control of its magnetization as well as a control of the polarization state by magnetic fields, $Ba_2Mg_2Fe_{12}O_{22}$ is interesting since magnetic fields of only 30 mT are sufficient to switch its spontaneous polarization. The latter system, cupric oxide CuO, was demonstrated to be multiferroic up to 230 K, proving that magnetically induced ferroelectricity is not restricted to low temperatures and can be considered as a quite robust feature. An extended list of related materials can be found in the review articles [29, 36] or [37].

However, the few examples given above perfectly reflect the diversity of magnetic structures, that can induce a ferroelectric polarization. Depending on the symmetry of each particular system, possible magnetic structures reach from simple Néel-like up-up-down-down [95] structures to complex cycloidal [77], elliptical [96] or transverse conical [26] spin arrangements.

A common but unfavorable feature of all these joint-order-parameter multiferroics is, that the magnetically induced spontaneous polarization is typically small, being in the order of 1 to 10 nC/cm^2. For a comparison, in conventional ferroelectrics a spontaneous polarization of $\gtrsim 10~\mu C/cm^2$ is observed. Nevertheless, it is the unique robustness of the magnetoelectric interaction in these systems rendering them interesting for future device applications.

The following section provides an overview of different models that explain ferroelectric behavior, arising as a consequence of magnetic ordering.

---

[2]Other terms used in the literature in this context are *improper* [91] or *pseudo-proper* [92] multiferroics, *sole-order-parameter* [93] multiferroics, or *type-II* [35] multiferroics.

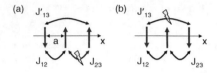

Figure 3.3: Frustration in a one-dimensional spin chain with antiferromagnetic nearest- (J_{ij}) and next-nearest-neighbor (J'_{ij}) exchange. Since both interactions are antiferromagnetic, they cannot be satisfied at the same time.

3.3   Magnetically-induced ferroelectricity

This section covers the fundamentals of ferroelectric order, forming as a direct consequence of magnetism. Although different review articles are devoted to this special class of magnetoelectric multiferroics, most of them merely provide a list of materials and observed effects [29, 32, 36]. Moreover, comprehensive textbooks that especially focus on this topic are completely missing.

Therefore, a rather extended discussion of the relevant aspects is given in the following. Starting point is the magnetism of systems with competing magnetic interactions.

3.3.1   Modulated Magnetic Ordering

The main magnetic interaction between localized spins $\mathbf{S}_{i,j}$ at site i and j is the exchange interaction J_{ij}. In the Heisenberg-model [97] it is described by the Hamiltonian

$$\mathcal{H} = \frac{1}{2} \sum_{i,j} J_{ij} \mathbf{S}_i \mathbf{S}_j \tag{3.2}$$

and induces different types of magnetic ordering, depending on the sign and distance dependence of J_{ij}. In the case of competing antiferromagnetic *nearest*-neighbor and *next-nearest*-neighbor interactions, denoted by $J_{ij} = J\delta_{i,i+1} > 0$ and $J'_{ij} = J'\delta_{i,i+2} > 0$, respectively, equation (3.2) becomes

$$\mathcal{H} = \sum_i J\mathbf{S}_i\mathbf{S}_{i+1} + \sum_i J'\mathbf{S}_i\mathbf{S}_{i+2} \,. \tag{3.3}$$

As depicted in figure 3.3, the competition immediately leads to frustration in a one-dimensional spin chain, because both interactions cannot be satisfied simultaneously.

To derive the magnetic ordering for competing J and J', the ground state of the Hamiltonian (3.3) has to be calculated. For this purpose, it is convenient to consider the Fourier transform of the general exchange interaction, since the Hamiltonian is invariant under translations $i \rightarrow i'$. The Fourier transform of a general J_{ij} is given by

$$\tilde{J}(k) = \sum_j J_{0j} e^{ikx_j} \,, \tag{3.4}$$

with x_j denoting the position of spin j in the chain [98]. In the case of nearest- (NN) and next-nearest-neighbor (NNN) exchange, equation (3.4) simplifies to

$$\tilde{J}(k) = \begin{cases} J_{01}(e^{ikx_{j-1}} + e^{ikx_{j+1}}) = 2J\cos ka & \text{, for NN interactions} \\ J'_{02}(e^{ikx_{j-2}} + e^{ikx_{j+2}}) = 2J'\cos 2ka & \text{, for NNN interactions} \end{cases} \tag{3.5}$$

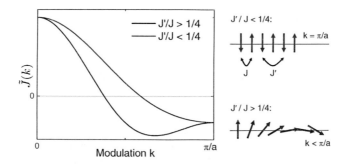

Figure 3.4: (a) Dependence of $\tilde{J}$ on k: Two qualitatively different solutions exist for a one-dimensional chain of spins with competing magnetic interactions, depending on the strength of the exchange coupling J and J'. The classical ground state can either be a simple antiferromagnetic arrangement of spins or a long-range modulation of the spin structure as sketched on the right hand side of the graph.

where $x_{j+1} = -x_{j-1} = a$ is the distance between neighboring spins and $J_{01} \equiv J$, $J_{02} \equiv J'$. Note that k in equation (3.4) determines the period of the modulated exchange interaction $\tilde{J}(k)$. Since the Hamiltonian (3.3) consists of terms $J\mathbf{S}_i\mathbf{S}_j$, only modulations $k = k_{min}$ will be realized that minimize $\tilde{J}(k)$ in order to minimize the energy of the system. In the present case k_{min} can be found by calculating the minimum of

$$\tilde{J}(k) = 2J\cos ka + 2J'\cos 2ka \qquad (3.6)$$

with respect to k:

$$\frac{d\tilde{J}(k)}{dk} = -2J\sin ka\left(1 + 4\frac{J'}{J}\cos ka\right) = 0 \qquad (3.7)$$

$$\Rightarrow \quad \begin{cases} k_{min} = \frac{\pi}{a} & \text{for } \frac{J'}{J} < \frac{1}{4} \\ k_{min} = \frac{1}{a}\arccos(-\frac{J}{4J'}) & \text{for } \frac{J'}{J} > \frac{1}{4} \end{cases} . \qquad (3.8)$$

The results are visualized in figure 3.4 showing $\tilde{J}(k)$ as function of k. One can see that two qualitatively different solutions k_{min} exist depending on the ratio $\frac{J'}{J}$:

1. For $\frac{J'}{J} < \frac{1}{4}$ the effective exchange interaction $\tilde{J}(k)$ reaches its minimum at $k_{min} = \frac{\pi}{a}$. As we will see in the following, this corresponds to a simple antiferromagnetic ground state of the spin system.

2. In contrast, an incommensurate modulation $0 < k_{min} < \frac{\pi}{a}$ is energetically favorable for $\frac{J'}{J} > \frac{1}{4}$. In this case, the classical ground state is a magnetic spiral [29].

However, the particular long-range order of the spins still has to be deduced. The simplest way is to consider a mean-field approximation of the Hamiltonian (3.2). The so-called mean-field approximation (3.9) leads to a decoupling of the spins, so that the spin $\mathbf{S}_i$ can be considered as interacting with an effective field generated by the other spins of the system. This leads to

$$\mathcal{H} = \sum_i \sum_j J_{ij}\langle\mathbf{S}_j\rangle\mathbf{S}_i , \qquad (3.9)$$

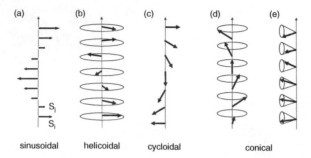

Figure 3.5: Schematic illustration of different modulated magnetic structures. (a) For a sinusoidal arrangement the local magnetization value changes in space. (b)-(e) In the case of helicoidal, cycloidal or conical spin structures the local magnetization value is constant whereas its orientation periodically changes.

and the effective field H_{eff} is given by

$$\sum_j J_{ij} \langle \mathbf{S}_j \rangle = g\mu_B H_{\text{eff}} . \tag{3.10}$$

Hence, the internal magnetization changes according to $\langle \mathbf{S}_j \rangle$, which again can be represented by its Fourier transform:

$$\langle \mathbf{S}_j \rangle = \int_{-\infty}^{+\infty} dk \; \langle \mathbf{S}(k) \rangle \, e^{ikx_j} = \langle \mathbf{S}(k_{\text{min}}) \rangle \, e^{ik_{\text{min}}x_j} . \tag{3.11}$$

Here it is already taken into account that only those values of $k = k_{\text{min}}$ occur that minimize the energy of the system. Since $\langle \mathbf{S}(k_{\text{min}}) \rangle \equiv \langle \mathbf{S} \rangle$ is constant, the variation of the effective field along the one-dimensional chain is determined by

$$\langle S_j^z \rangle = \langle S^z \rangle \, e^{ik_{\text{min}}x_j} , \tag{3.12}$$

if z is chosen to be the quantization axis of the spin. In the most general case, equation (3.12) describes a spin-spiral structure with period $\lambda = \frac{2\pi}{k_{\text{min}}}$.
For $k_{\text{min}} = \frac{\pi}{a}$ equation (3.12) represents a simple $\uparrow\downarrow\uparrow\downarrow$ antiferromagnetic ordering along the chain with $\pm\langle S_j^z \rangle$, depicted in figure 3.4 ($x_j = \pm na$ with $n \in \mathbb{N}$).
In general, the analysis of the spin structure has to be carried out more carefully, because different structures may exist with period λ. One example is a sinusoidal spin-density wave with a magnetization at each site parallel to the z axis. A sinusoidal spin arrangement is described by $\langle S_j^z \rangle = \langle S \rangle \cos(k_{\text{min}}x_j)$ and is displayed in figure 3.5(a). Alternatively, helicoidal structures with $\langle S_j^x \rangle = \langle S \rangle \cos(k_{\text{min}}x_j)$ and $\langle S_j^z \rangle = \langle S \rangle \sin(k_{\text{min}}x_j)$ are possible. Here, the local magnetization value at each site is constant, while its orientation rotates in the xz plane (see figure 3.5(b)). Further modulated spin structures that typically arise due to competing magnetic interactions are presented in figure 3.5 (c) to (e).
Note that the magnetic order parameters η of the above discussed periodically modulated structures are typical examples for order parameters which are related to a nonzero wave vector $\mathbf{k}$. This has to be taken into account in their group theoretical treatment as it was discussed in the previous section 2.2.1.

3.3.2 Ionic polarization by symmetric exchange interactions

After showing in a very general way that modulated magnetic ordering can inherently occur in magnetically frustrated systems, we now discuss in detail how magnetism can induce a macroscopic polarization $\mathbf{P}$. Depending on the symmetry of a magnetically ordered crystal, different mechanisms can be responsible for a polar displacement. Such displacements occur when exchange interactions are present that favor a deformation of the local structure in order to minimize the total energy E of the system.

First, the so-called exchange-striction model is introduced, which is bases on the *symmetric* Heisenberg exchange interaction of neighboring spins $\propto \mathbf{S}_i\mathbf{S}_{i+1}$ (see equation (3.2)).

In this context, the Goodenough-Kanamori-Anderson (GKA) rules play an important role [99]. The GKA rules indicate whether ferro- or antiferromagnetic ordering occurs when positive magnetic ions interact via an intermediary non-magnetic anion, called superexchange.

As an example, we consider the superexchange of two manganese ions (Mn^{3+}) via an intermediary oxygen anion (O^{2-}). Here, the local magnetic moment $S = 3/2$ is formed by three electrons occupying the d_{xy}, the d_{xz}, and the d_{yz} orbital (t_{2g} electrons) of Mn^{3+}.

A fourth electron occupies the d_z orbital (e_g electron) that partially overlaps with the p orbital of the O^{2-} ion, leading to virtual hopping. Virtual hopping means, that no "real" charge transfer occurs, but the probability to find the e_g electron in an orbital of the neighboring oxygen is nonzero, due to the overlap of the associated wave functions.

According to the GKA rules, the superexchange leads to antiferromagnetic ordering if the Mn–O–Mn bond angle is close to 180° which is illustrated in figure 3.6(a). On the contrary, a bond angle of 90° causes ferromagnetic coupling (figure 3.6(b)).

Concerning a polar displacement in a magnetically ordered system we learn from the GKA rules, that *antiparallel* spins lead to an *increase* of the Mn–O–Mn bond angle, whereas *parallel* spins favor a *decrease* of the bond angle, compared to the paramagnetic state (see figure 3.6(c)).

Thus, for fixed positions of the Mn^{3+} ions the superexchange results in a relaxation of the O^{2-} ion in the magnetically ordered state, potentially leading to $\mathbf{P} \neq 0$.

Whether a macroscopic polarization shows up or not depends on the symmetry of the magnetically ordered system. As shown in the upper part of figure 3.6(d), local distortions of the lattice alternate for a simple up-down-up-down spin arrangement. Apparently, the polar displacements globally compensate and no macroscopic net polarization evolves – the inversion symmetry of the system is conserved in the magnetically ordered phase.

Only when the magnetically ordered system lacks inversion symmetry a macroscopic polarization shows up [29]. One example is depicted in the lower part of figure 3.6(d). In this case the up-up-down-down ordering of the spins generates polar displacements of the intermediary anions that violates the center of inversion. Due to the different Mn–O–Mn bond angles of antiferro- and ferromagnetically coupled neighbors, no cancelation of $\mathbf{P}$ occurs.

Examples for magnetically induced ferroelectricity, mainly caused by a symmetric exchange interaction can be found in the RMn_2O_5 series (R = Y, Tb, Er,...) [40, 100, 101]. Another well studied system with mostly collinear antiferromagnetic spins is orthorhombic $HoMnO_3$ [95].

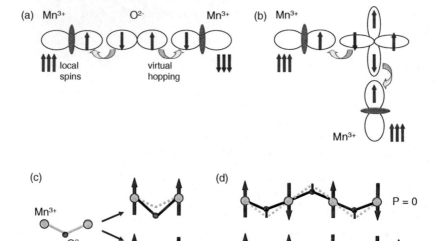

Figure 3.6: Magnetic ordering due to superexchange. (a) According the the GKA rules, a Mn–O–Mn bond angle of 180° leads to antiferromagnetic ordering of the local Mn^{3+} moments. (b) In contrast, a ferromagnetic coupling of the local moments is energetically favorable for a Mn–O–Mn bond angle of 90°. (c) Ionic displacements caused by symmetric exchange interactions. While parallel ordered spins favor a decrease of the Mn–O–Mn bond angle, antiferromagnetic exchange leads to an increase. (d) For simple antiferromagnetic ordering, local polar displacements compensate. Only when the magnetically induced ionic displacements break the spatial inversion symmetry a macroscopic polarization is observed.

3.3.3   Ionic polarization by antisymmetric exchange interactions

So far we considered symmetric exchange interactions only. In magnetic crystals with low symmetry, the exchange interaction can also contain an anisotropic term due to spin-orbit coupling [102]. The resulting Hamiltonian is

$$\mathcal{H} = \frac{1}{2}\sum_{i,j} J_{ij}\mathbf{S}_i\mathbf{S}_j + \underbrace{\frac{1}{2}\sum_{i,j} \mathbf{D}_{ij}\cdot(\mathbf{S}_i\times\mathbf{S}_j)}_{=\mathcal{H}_{\mathrm{DM}}} \; . \tag{3.13}$$

The second term $\mathcal{H}_{\mathrm{DM}}$ is called the Dzyaloshinskii-Moriya interaction (DMI), with the Dzyaloshinskii-Moriya vector $\mathbf{D}_{ij}$ ($\mathbf{D}_{ij} = -\mathbf{D}_{ji}$).
Regarding again two transition metal ions (Mn^{3+}) with an intermediary oxygen ion (O^{2-}), the direction of $\mathbf{D}_{ij}$ depends on the position of the oxygen ion and can be denoted by [29]

$$\mathbf{D}_{ij} \propto \mathbf{x} \times \mathbf{r}_{ij} \; ,$$

where $\mathbf{r}_{ij}$ is a unit vector connecting neighboring magnetic ions and $\mathbf{x}$ parametrizes the oxygen displacement as sketched in figure 3.7(a).

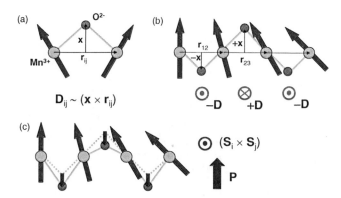

Figure 3.7: Ionic displacements caused by antisymmetric exchange interactions. (a) Relation between the Dzyaloshinskii-vector D_{ij} and the shift of oxygen ions x due to canted Mn^{3+} spins. (b) The Dzyaloshinskii-vector alternates in sign for zig-zag chains of Mn^{3+} and intermediary O^{2-} (c) Relaxation of the O^{2-} ions according to equation (3.13) leads to a macroscopic polarization.

Opposite to the symmetric exchange interaction, the DMI favors a noncollinear ordering of neighboring spins, naturally stabilizing two- or three dimensional helicoidal, cycloidal or conical spin structures as depicted in figure 3.5(b) to (e).

Usually, a structural anisotropy as displayed in figure 3.7(a) gives rise to the DMI interaction which results in a canting of the spins. Magnetically-induced ferroelectricity is the "inverse" effect and, therefore, often referred to as inverse DMI in the literature [103]. Here, noncollinear spins cause a nonzero term $\mathcal{H}_{DM}$ and thus induce a polar displacement.

The presence of antisymmetric exchange interactions for instance explains why cycloidal magnetic ordering can generate a macroscopic polarization. As figure 3.7(b) shows, D_{ij} alternates along a zig-zag chain of Mn^{3+} ions and intermediary O^{2-} ions, while $S_i \times S_j$ remains constant.

According to equation (3.13), it is energetically favorable to enhance the oxygen displacement x for $D_{ij} < 0$, whereas a decrease of x takes place for $D_{ij} > 0$. In total, this leads to a collective movement of all anions in the same direction relative to the fixed cations. As pointed out in figure 3.7(c), this corresponds to a spontaneous polarization P, perpendicular to the propagation direction of the spin cycloid r_{ij} and its rotation axis $S_i \times S_j$.

General relation between the direction of P and noncollinear spins: Ionic displacements caused by Dzyloshinskii-Moriya-type interactions were first proposed by Sergienko *et al.* to be the reason for ferroelectricity in spin-spiral ferroelectrics [103]. Independently, Mostovoy derived a general formula, describing the relation between the induced spontaneous polarization P and the ordered spins, being valid in the presence of Dzyaloshinskii-Moriya interactions [31]. The relation is given by

$$P \propto \delta\chi_e(S_i \times S_j) \times Q \,, \tag{3.14}$$

where $\mathbf{Q}\|\mathbf{r}_{ij}$ is the propagation vector of the modulated spin structure. The coupling constant δ takes into account the strength of the coupling between the magnetic and the electric ordering and χ_e denotes the electric susceptibility. The term $\mathbf{C} := \mathbf{S}_i \times \mathbf{S}_j$ identifies the rotation axis of the spin arrangement and often is referred to as *chirality* regarding periodically modulated spins.

The relation (3.14) points out the importance of noncollinear spins for magnetically induced ferroelectricity and describes experimentally observed correlations between $\mathbf{P}$ and $\mathbf{S}_i \times \mathbf{S}_j$ in many systems as $TbMnO_3$, $DyMnO_3$ or $Ni_3V_2O_8$ [11,12,77].

However, being derived in the framework of a continuous Landau theory, relation (3.14) includes no microscopic properties at all. It is derived by analyzing the lowest order-coupling terms between a spontaneous polarization $\mathbf{P}$ and Lifschitz invariants $\mathbf{M}(\Delta\mathbf{M})$, being related to the magnetization $\mathbf{M}$ of the system[3]. Such Lifschitz invariants are allowed to contribute to the free energy of a system whenever the inversion symmetry is violated [104,105].

Hence, equation (3.14) is generally valid for non-centrosymmetric systems, independent of the microscopic background.

3.3.4 Electronic polarization by spin supercurrent

Up to now, we considered magnetically induced ferroelectricity caused by the relaxation of ions in a non-centrosymmetric environment. Alternatively, the spontaneous polarization may stem from electronic contributions, while the ions stay at their centrosymmetric positions [91]. In this case one speaks of a purely electronic polarization.

A microscopic explanation for electronic polarizations is given by the so-called Katsura-Nagaosa-Balatsky model (KNB model). The KNB model shows that a spontaneous polarization of electronic origin can arise in the case of neighboring noncollinear spins $\mathbf{S}_i$ and $\mathbf{S}_j$ due to a nonzero spin supercurrent $\mathbf{j}_s^{eq}$ (also equilibrium spin current) defined by [30,106]

$$\mathbf{j}_s^{eq} = \frac{J_{ij}}{\hbar}\mathbf{S}_i \times \mathbf{S}_j \; . \tag{3.15}$$

Here, J_{ij} is the Heisenberg coupling constant (see equation (3.2)). The mechanism leading to spin supercurrents is basically the same as for the superexchange, i.e. virtual hopping of electrons, whereas in the present case also the spin of an electron is taken into account. The relation (3.15) is illustrated in figure 3.8(a) for a centrosymmetric arrangement of two Mn^{3+} ions and an intermediary O^{2-} ion.

Apparently, the direction of $\mathbf{j}_s^{eq}$ is constant in the case of cycloidal magnetic ordering, being depicted in figure 3.8(b). As discussed in detail in reference [30], this leads to a macroscopic spontaneous polarization of electronic origin

$$\mathbf{P} \propto \mathbf{r}_{ij} \times \mathbf{j}_s^{eq} \propto \mathbf{r}_{ij} \times (\mathbf{S}_i \times \mathbf{S}_j) \; . \tag{3.16}$$

Note that interactions of Dzyaloshinskii-Moriya-type vanish in the current example, because ions are assumed to be fixed to their centrosymmetric positions ($\mathbf{x} = 0 \Rightarrow \mathbf{D}_{ij} = 0$). Nevertheless, equation (3.16) is quite similar to equation (3.14), which has been discussed above with respect to ionic displacements. This follows from the fact that equation (3.14) is derived from phenomenological considerations and does not include a microscopic model. Therefore, equation (3.14) applies to ionic as well as electronic polarization contributions.

---

[3]Since a continuous approach is used, the magnetization $\mathbf{M}$ is considered instead of spins $\mathbf{S}_i$.

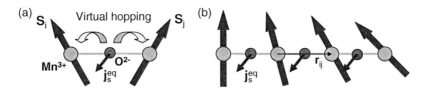

Figure 3.8: (a) Ferroelectricity due to spin current: According to the BNK model, ferroelectricity can arise from noncollinear neighboring spins when the intermediary anion is pinned in its centrosymmetric position. In this case no interactions of Dzyaloshinskii-Moriya type are allowed (see also figure 3.7). Here, the spontaneous polarization originates from a nonzero spin supercurrent $\mathbf{j}_s^{eq}$. (b) The direction of $\mathbf{j}_s^{eq}$ is the same for all pairs of magnetic ions considering cycloidal magnetic order which leads to a macroscopic polarization of electronic origin.

Separation of ionic and electronic contributions to P: To identify whether ionic displacements, i.e. a *lattice mechanism*, electronic charge redistributions, or a combination of both causes the spontaneous polarization in joint-order-parameter multiferroics is a challenging experimental task.

The difficulties follow from the small polarization values of these systems, being in the order of 1 to 10 nC/cm². Such a small ferroelectric polarization would already arise due to ionic displacement in the order of 10^{-3} Å, which is at the resolution limit of neutron, X-ray, and Raman experiments. Thus, it often cannot be said if intermediary anions are shifted with respect to their centrosymmetric position in phases with $\mathbf{P} \neq 0$ or not [101, 107–113]. Present theories mostly focus on orthorhombic $R\mathrm{MnO_3}$ and $R\mathrm{Mn_2O_5}$. Some support a predominantly ionic [103, 107] or predominantly electronic [30, 114, 115] ferroelectricity while others consider comparable influence of the two [95, 116, 117].

Hence, there is an intense and controverse debate about the microscopy of magnetically induced ferroelectrics.

3.4 Domains

Ferroic materials possess two or more energetically equivalent orientation states regarding their order parameter [48, 70]. Such orientation states, i.e. volumes of the same homogeneous crystalline structure but different spatial orientations of the order parameter, are called *domains*. The domains are separated by *domain walls*. Within or at these domain walls the direction of the order patameter η changes [118].

The formation of domains is an essential feature of any ferroic system. By breaking up into domains, conventional ferromagnets or ferroelectrics can substantially reduce their magnetostatic or electrostatic energy. Thus, provided the decrease in energy is greater than the energy needed to form domain walls, a multi-domain state will arise in the ordered phase [57, 119, 120].

However, due to the lowering of the symmetry at the paramagnetic → antiferromagnetic phase transition, also antiferromagnets tend to form domains, although the presence of antiferromagnetic domains does not lower the free energy [121].

Explicitly complex domain structures arise in split-order-parameter multiferroices, where magnetic and electric domains coexist [10, 72]. To what extend the concept of magnetic and electric domains applies to magnetically induced ferroelectrics (joint-order-parameter

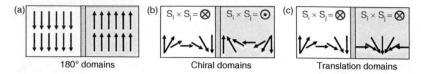

Figure 3.9: Different types of domains in ferroic materials. (a) Adjacent domains of opposite order parameter η are called 180° domains. (b) Periodically modulated magnetic structures with $\mathbf{S}_i \times \mathbf{S}_j \neq 0$ potentially form domains of opposite chirality $\pm|\mathbf{S}_i \times \mathbf{S}_j|$. (c) Discontinuities in the periodicity of a modulated spin structure lead to the formation of translation domains.

multiferroics) used to be an entirely open question and therefore has been investigated in detail in the context of this thesis.

Figure 3.9 exemplifies three different domain states that might play a role in magnetically induced ferroelectrics, namely 180° domains, chiral domains, and translation domains. One speaks of 180° *domains*, whenever adjacent domains have opposite magnetic or ferroelectric order parameters $\pm\eta$ as depicted in figure 3.9(a).

Figure 3.9(b) and (c) show special cases of magnetic domains. So-called *chiral domains* occur for example in helical antiferromagnets as Dysprosium or Terbium [122,123]. In both materials, the chiral domains consist of helices of opposite chirality $\mathbf{C} := (\mathbf{S}_i \times \mathbf{S}_j)$.

In general, chiral domains can arise whenever periodically modulated spin structures exhibit a nonzero chirality $\mathbf{S}_i \times \mathbf{S}_j \neq 0$. In figure 3.9(b) the chiral domains of a system with cycloidal magnetic order are schematically depicted.

Another type of domains that may occur for periodically modulated spin structures are *translation domains*. Translation domains naturally evolve when isolated *islands* (nucleation centers) of an ordered phase grow independently and finally converge. Thus, each domain exhibits a continuously modulated spin structure, whereas domain walls in between translation domains correspond to a discontinuity in the phase of the modulation. Figure 3.9(c) illustrates translation domains for homochiral ($\mathbf{C} = $ const) cycloidal magnetic order.

Chapter 4

Experimental method

This chapter covers the experimental methods used for the investigation of magnetically induced ferroelectrics. All measurement were performed by nonlinear optics — a technique that is explicitly powerful regarding the investigation of complex ferroic systems with coexisting order parameters [124].

In the first sections of this chapter the general principles of nonlinear optics are reviewed. The theoretical overlook is followed by a description of the experimental setup. Part of this thesis was to set up and equip two new workplaces for nonlinear optical experiments. Their specific scopes of application are briefly outlined.

In the last part, different applied nonlinear optical techniques are discussed. In the context of this thesis these techniques have been proved feasible regarding the investigation of *incommensurate* antiferromagnetic long-range order. Special emphasis is put on spatially-resolved measurements. For the first time, spatially-resolved nonlinear optical measurements have been applied for imaging of the complex domain topology in systems with periodically modulated spin structures, i.e. magnetically induced ferroelectrics.

4.1 Nonlinear optics

In conventional optics only linear interactions of light and matter are considered and the relation between components of an applied light field $\tilde{\mathbf{E}}(t)$ and the components of the induced electric-diplole moment $\tilde{\mathbf{P}}(t)$ is given by[1]

$$\tilde{P}_i(t) = \epsilon_0 \sum_{j=1}^{3} \chi_{ij} \tilde{E}_j(t) \,, \tag{4.1}$$

with $\hat{\chi}$ as linear susceptibility tensor. However, this linear approximation is no longer valid when strong light fields give rise to nonlinear interactions of light and matter. Often the nonlinear response can be described by a power series in the field strength $\tilde{E}(t)$ [125, 126]

$$
\begin{aligned}
\tilde{P}_i(t) &= \epsilon_0 \sum_{j=1}^{3} \chi_{ij} \tilde{E}_j(t) + \epsilon_0 \sum_{j,k=1}^{3} \chi_{ijk} \tilde{E}_j(t)\tilde{E}_k(t) + \epsilon_0 \sum_{j,k,l=1}^{3} \chi_{ijkl} \tilde{E}_j(t)\tilde{E}_k(t)\tilde{E}_l(t) + \cdots \\
&= \tilde{P}_i^{(1)}(t) + \tilde{P}_i^{(2)}(t) + \tilde{P}_i^{(3)}(t) + \cdots \,,
\end{aligned}
\tag{4.2}
$$

---

[1]Here, the notation $\tilde{\mathbf{P}}(t)$ is used to emphasize that the polarization induced by the light field $\tilde{\mathbf{E}}(t)$ rapidly varies in time. Constant properties as a ferroelectric polarization are written without the tilde.

i.e. a generalization of equation (4.1). Accordingly, χ_{ijk} and χ_{ijkl} are components of the second- and third-order nonlinear optical susceptibility tensors, respectively.

4.1.1   Second harmonic generation

We now turn to a special case of nonlinear response. Regarding a monochromatic plane wave, the electric-field strength in the power series (4.2) can be represented by

$$\tilde{\mathbf{E}}(\omega, t) = \tilde{\mathbf{E}} e^{i\omega t} \ . \tag{4.3}$$

This implies $\tilde{\mathbf{P}}^{(2)}(t) \propto e^{i2\omega t}$ for the second-order polarization, indicating its dependence on the second harmonic frequency 2ω. The associated physical process is called optical *second harmonic generation*. It describes the simultaneous absorption of two photons at frequency ω, inducing an electric-dipole (ED) oscillation at double frequency 2ω. Excitation and response are linked by the nonlinear susceptibility tensor $\hat{\chi}^{\mathrm{ED}}$, so that the induced electric-dipole moment can be written as

$$\tilde{P}_i(2\omega) = \epsilon_0 \sum_{j,k=1}^{3} \chi_{ijk}^{\mathrm{ED}}(2\omega) \tilde{E}_j(\omega) \tilde{E}_k(\omega) \ . \tag{4.4}$$

This excited electric-dipole moment again is the source of an electromagnetic wave, constituting the SHG signal. Note that the indices j, k express the polarization direction of the incident light fields, while i determines the direction of the induced electric-dipole oscillation.

The leading higher-order multipole moments that potentially contribute to an emitted light wave are the magnetic-dipole moment

$$\tilde{M}_i(2\omega) = \frac{c}{\epsilon_0 n(\omega)} \sum_{j,k=1}^{3} \chi_{ijk}^{\mathrm{MD}}(2\omega) \tilde{E}_j(\omega) \tilde{E}_k(\omega) \tag{4.5}$$

and the electric-quadrupole moment

$$\tilde{Q}_{ij}(2\omega) = \frac{\epsilon_0 c}{2i\omega n(\omega)} \sum_{k,l=1}^{3} \chi_{ijkl}^{\mathrm{EQ}}(2\omega) \tilde{E}_k(\omega) \tilde{E}_l(\omega) \tag{4.6}$$

with $\hat{\chi}^{\mathrm{MD}}$ and $\hat{\chi}^{\mathrm{EQ}}$ as nonlinear susceptibilities, while $n(\omega)$ is the refractive index of the material at frequency ω. Hence, the complete related source term for SHG light is given by [124, 127]

$$\tilde{\mathbf{S}} = \mu_0 \frac{\partial^2 \tilde{\mathbf{P}}}{\partial t^2} + \mu_0 \left(\nabla \times \frac{\partial \tilde{\mathbf{M}}}{\partial t} \right) - \mu_0 \left(\nabla \frac{\partial^2 \hat{Q}}{\partial t^2} \right) \ . \tag{4.7}$$

The ED term $\propto \tilde{\mathbf{P}}$ is the leading contribution to $\tilde{\mathbf{S}}$. It exceeds the magnetic-dipole (MD) term $\propto \tilde{\mathbf{M}}$ and the electric-quadrupole (EQ) term $\hat{Q}$ by a factor $\lambda/2\pi a \approx 100$, where λ is the wavelength of the light and a the lattice constant of the crystal.

A schematic illustration of the SHG process is given in figure 4.1. Red arrows indicate the two absorbed photons, while the green arrows represent the emitted light wave. The illustration furthermore points out, that a decomposition of different SHG contributions is possible. For instance SHG of crystallographic origin can superimpose with SHG, caused by any kind of ferroic ordering.

To separate such signals, a symmetry- and temperature-dependent analysis of the detected signals is necessary. For this purpose the relation between SHG contributions and the physical properties of a system is discussed in the following section 4.2.

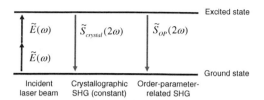

Figure 4.1: Schematic illustration of the SHG process. The system is excited by simultaneous absorption of two photons at frequency ω of an intense light field, i.e. a strong laser. The nonlinear response of the system at frequency 2ω can be separated in purely crystallographic and/or order-parameter-related contributions, denoted by $\hat{S}_{\mathrm{crystal}}$ and $\hat{S}_{\mathrm{OP}}$, respectively.

4.2 Nonlinear susceptibilities

4.2.1 Symmetry dependence

The multipole moments in equations (4.4), (4.5), and (4.6) represent the response of a material in the presence of an intense light field. One can see that the response of any system depends on its nonlinear susceptibility $\hat{\chi}$. The tensor $\hat{\chi}$ reflects the physical properties of the material and therefore is called a *property tensor*. All tensor components are determined by the symmetry of the system and so are possible SHG contributions.

In general, this is known as *Neumann's principle*, stating that any type of symmetry corresponding to the point group of the crystal is possessed by every physical property of the crystal [47].

Thus, to determine the components of $\hat{\chi}$, its transformation behavior under the point-symmetry operations of the crystal has to be investigated. In the end, this analysis provides information about the relation between detected SHG signals and the symmetry of a crystal, or even more important, the symmetry of present order parameters.

The procedure is quite similar to the one adapted for specifying the form of the free energy in section 2.1.2. However, there is an essential difference: While in the case of the free energy the transformation behavior of polynomials was considered, we now have to take into account the specific transformation behavior of tensors.

Basically, tensors are classified by their transformation behavior under the two fundamental parity operations, i.e. space inversion and time inversion, both having eigenvalues ± 1. The two related operators in the context of tensors are typically denoted by $\hat{I}$ ($\mathbf{r} \rightarrow -\mathbf{r}$) and $\hat{T}$ ($t \rightarrow -t$), respectively.

Polar and axial tensors: Concerning the spatial inversion, we distinguish between polar and axial tensors. Decisive for the classification is, according to which of the following equations a tensor $\hat{d}^{(n)}$ of rank n transforms under $\hat{I}$:

$$\text{polar}: \quad \hat{I}\hat{d}^{(n)}(\mathbf{r}, t) = (-1)^n \; \hat{d}^{(n)}(-\mathbf{r}, t) \tag{4.8}$$

$$\text{axial}: \quad \hat{I}\hat{d}^{(n)}(\mathbf{r}, t) = (-1)^{n+1}\hat{d}^{(n)}(-\mathbf{r}, t) \tag{4.9}$$

The definition implies important general consequences: In a centrosymmetric crystal, all axial tensors of even rank and all polar tensors of odd rank have to vanish identically. Moreover, the product of two polar or two axial tensors transforms like a polar tensor, whereas the product of one polar and one axial tensor behaves like an axial tensor.

Regarding the nonlinear susceptibilities of the multipoles in equations (4.4) to (4.6), we can conclude that electric-dipole contributions to a SHG signal are only allowed in non-centrosymmetric crystals, because $\hat{\chi}^{\text{ED}}$ is a polar tensor of odd rank. In contrast, the magnetic-dipole and the electric-quadrupole contributions are allowed when a center of inversion is present, because $\hat{\chi}^{\text{MD}}$ is an axial tensor of odd rank and $\hat{\chi}^{\text{EQ}}$ is a polar tensor of even rank.

In section 2.1.2 the example of a ferroelectric phase transition in a crystal with point-group symmetry $4/mmm$ has been discussed. In that case, electric-dipole contributions are forbidden by symmetry in the high-temperature paraelectric phase, because spatial inversion is a symmetry operation of $4/mmm$. Only in the ferroelectric phase the center of inversion is violated by a polar displacement and nonzero components χ^{ED}_{ijk} become symmetry-allowed. Thus, in a temperature-dependent measurement the onset of additional SHG contributions at T_c would indicate the ferroelectric phase transition.

Tensors of i- and c-type: The classification as i- or c-tensor reflects either the invariance (i-) or the change (c-) of the sign of a tensor under time reversal:

$$\text{i-tensor}: \qquad \hat{T}\hat{d}^{(n)}(\mathbf{r}, t) \;=\; +\hat{d}^{(n)}(\mathbf{r}, -t) \qquad\qquad (4.10)$$

$$\text{c-tensor}: \qquad \hat{T}\hat{d}^{(n)}(\mathbf{r}, t) \;=\; -\hat{d}^{(n)}(\mathbf{r}, -t) \qquad\qquad (4.11)$$

The two equations indicate that a product of two i- or two c- tensors always transforms like an i-tensor (is invariant under $\hat{T}$), whereas the product of an i- and a c-tensor changes sign under the time-reversal operation.

The concept of i- and c-tensors is important for magnetically ordered systems. The onset of magnetic ordering generally violates the time-inversion symmtery of a crystal, allowing for nonzero c-tensors. Again, this will affect the SHG-yield and can be detected in the experiment.

4.2.2   Order-parameter dependence

It was just shown, that property tensors are determined by the point group of a material, which has to change at a ferroic phase transition (see section 2.1.1). We now discuss the dependence of the SHG susceptibility on the order parameter that induces the transition at T_c. One can show by using a generalized Ginzburg-Landau theory [128] that all nonzero components of $\hat{\chi}^{(T<T_c)}$ in the ordered phase can be obtained from the symmetry of the susceptibility tensor $\hat{\chi}^{(T>T_c)}$ in the high-temperature phase and the symmetry of the order parameter tensor $\hat{\eta}$. The general relation is given by

$$\hat{\chi}^{(T<T_c)} = \hat{\chi}^{(T>T_c)}_0 + \hat{\chi}^{(T>T_c)}_1 \hat{\eta} + \hat{\chi}^{(T>T_c)}_2 \hat{\eta}\hat{\eta} + \cdots , \qquad\qquad (4.12)$$

where the $\hat{\chi}^{(T>T_c)}_i$ denote the coefficients of the series, each being a tensor. Nonzero tensor components of $\hat{\chi}^{(T>T_c)}_i$ are determined by the symmetry of the crystal in its high-temperature phase above T_c. Note that the tensors $\hat{\chi}^{(T>T_c)}_i$ are not affected by the transition – only the order parameter $\hat{\eta}$ changes at T_c.

Equation (4.12) leads to the important conclusion that right below a ferroic phase transition the changes in the nonlinear susceptibility are proportional to the order parameter $\hat{\eta}$. This is valid, because the components of $\hat{\eta}$ are small in the vicinity of the phase transition and quadratic terms in the order parameter can be neglected.

Whenever two ferroic order parameters $\hat{\eta}_1$ and $\hat{\eta}_2$ arise at subsequent phase transitions ($T_1 > T_2$), the expansion (4.12) includes $\hat{\eta}_1$ and $\hat{\eta}_2$. This leads to contributions that are coupled to both order parameters. Taking into accout only a linear coupling to each of the order parameters $\hat{\eta}_i$, we get

$$
\begin{aligned}
\hat{\chi} &= \hat{\chi}_0^{(T > T_2)} + \hat{\chi}_1^{(T > T_2)} \hat{\eta}_2 \\
&= \hat{\chi}_{00}^{(T > T_1)} + \hat{\chi}_{01}^{(T > T_1)} \hat{\eta}_1 + \hat{\chi}_{10}^{(T > T_1)} \hat{\eta}_2 + \hat{\chi}_{11}^{(T > T_1)} \hat{\eta}_1 \hat{\eta}_2 .
\end{aligned}
\tag{4.13}
$$

All terms in the expansion exhibit a different dependence on the order parameters [129]. Obviously, nonzero terms depending on $\hat{\eta}_1$ and $\hat{\eta}_2$ are restricted to the low-temperature phase ($T < T_2$).

In summary, equation (4.13) expresses how ferroic order parameters and nonlinear suceptibilities are linked. Together with the representation of the multipole moments given by equations (4.4) to (4.6) and the source term (4.7), we can now understand the sensitivity of SHG measurement towards ferroic ordering. Moreover, the theoretical framework enables the prediction of allowed SHG contributions and their scaling behavior in the presence of ferroic order parameters.

4.3 Experimental setup

An elementary part of the preparative work for this thesis was to set up and start running two different workplaces for nonlinear optical experiments in two new laboratories. Both workplaces are typically used for measurements in transmission geometry, while each of the two laboratories has been designed with respect to a different scope of application.

A first workplace, namely "Nano1", is configured for nonlinear spectroscopy measurements, whereas the second workplace "Nano2" is optimized for domain imaging.

In this section, a brief overview is given about the transmission setup and the main differences regarding the equipment of the two laboratories Nano1 and Nano2 are specified.

4.3.1 Transmission geometry

The setup used for nonlinear optical measurements in transmission geometry is schematically depicted in figure 4.2(a).

The frequency-tripled beam of a Nd:YAG laser (355 nm) pumps a so-called optical parametric oscillator (OPO), serving as light source for the experiment [130]. The wavelength of the emitted light field of the OPO can typically be tuned in between ≈ 400 nm and ≈ 3000 nm (3.1 to 0.4 eV), so that nonlinear spectroscopy in the range of 6.2 to 0.8 eV becomes possible.

Behind the OPO the polarization direction of the linearly polarized fundamental beam is defined. This is realized by a Glan prism which is followed by a rotatable $\frac{\lambda}{2}$-waveplate denoted as polarizer in figure 4.2(a). Together with the analyzer (also rotatable) the configuration enables polarization-dependent measurements. The sample itself is mounted between polarizer and analyzer and optical filters behind the sample ensure that only the frequency-doubled SHG light is detected, while the fundamental beam is blocked.

For spatially resolved measurements a liquid nitrogen cooled CCD camera is used, where either a CCD camera or a photomultiplier tube can be used as a detector for spectroscopy measurements.

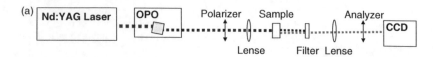

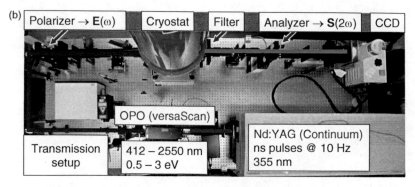

Figure 4.2: (a) Simplified standard setup for SHG measurements in transmission geometry. The combination of a Nd:YAG laser and an optical parametric oscillator (OPO) serves as tunable light source for the experiment. The polarization direction of the fundamental beam is set by the polarizer, i.e. a Glan prism followed by a rotatable $\frac{\lambda}{2}$-waveplate. A focussing lense allows to control the energy density at the sample position. Behind the sample optical filters block the fundamental wave, while frequency-doubled SHG signals can pass. Polarization-dependent detection of SHG contributions is enabled by the rotatable analyzer in front of the CCD camera. Additional optical components are needed to control the beam path and to image the sample (not shown). (b) Topview on the optical table in the laboratory "Nano2". The photo shows how the transmission setup (a) is realized at this workplace.

Figure 4.2(b) shows a topview on the optical table situated in Nano2, nicely visualizing how the transmission setup is realized in this laboratory. The setup in Nano1 is basically the same — the specific fields of application considering Nano1 and Nano2 result from the used pump laser and the OPO only.

Nano1: The workplace in the laboratory Nano1 is equipped with a nanosecond Nd:YAG laser (INFINITY) built by Coherent. This laser pumps the OPO system called NORMA by GWU-Lasertechnik. Here, the laser pulses have a length of ≈ 3 ns and the pulse frequency can be continuously varied between 0.1 and 100 Hz. Starting with a pump energy of 120 mJ (355 nm), the energy of the fundamental wave at the position of the sample is about 5 mJ, due the limited conversion efficiency of the OPO and reflection/absorption by optical components.
Furthermore, a seeding option of the OPO allows for reducing its spectral width from several meV to 20-50 μeV, predestining this setup for high-resolution spectroscopy [129].

Nano2: The workplace in the second laboratory Nano2 is constructed to provide rather high energies. In this case a Nd:YAG nanosecond laser (Powerlite Precision II) built by Continuum is the primary light source. At a fixed repetition rate of 10 Hz, laser pulses

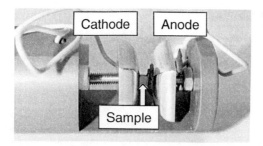

Figure 4.3: Sample holder designed for electric-field dependent SHG measurements. Adjustable polished non-magnetic electrodes ensure a homogeneous electric-field and enable the investigation of samples with arbitrary geometry in electric fields. The design allows for application in the Janis, as well as in the Oxford cryostat (see text).

with a length of 5-7 ns are emitted. The pulses have an energy of 300 mJ at 355 nm. In this case, a high-power OPO (versaScan by GWU-Lasertechnik) has been selected, providing an energy of about 50 mJ at the position of the sample.

The high intensities allow for instance to widen the diameter of the fundamental wave, so that the probed region in spatially-resolved measurements can be increased significantly.

4.3.2   Field- and temperature-dependent measurements

The investigation of physical effects in materials with strongly correlated electrons often requires measurements at low temperature. Especially for the study of magnetically induced ferroelectrics the application of cryogenics is essential, because these systems typically exhibit magnetic ordering well below room temperature.

For the low-temperature measurements presented in this thesis, two different types of optical cryostats equipped with quartz windows were used:

For zero-field measurements the Janis SVT-400 model was employed. In this variable temperature cryostat the sample is cooled by controlling the temperature of flowing helium vapor, so that temperatures between 1.4 and 300 K are accessible. The precision of the sample temperature is provided by an automatic temperature controller (LakeShore, Model 331) and specific thermometry (± 0.1 K).

Magnetic-field dependent measurements were performed using the Spectromag system of Oxford Instruments, including a split-pair superconducting magnet with a central field of 7 T at 4.2 K (8 T at 2.2 K). In its variable temperature inset (VTI) the sample is cooled by helium vapor, offering a temperature range of 1.5 to 300 K. The temperature is controlled automatically with an accuracy of ± 0.1 K.

To enable the electric-field dependent measurements presented in this thesis, an appropriate sample holder has been designed that fits into both cryogenic systems. Highly-polished surfaces of the non-magnetic electrodes help to avoid field peaks and ensure a homogenous electric field over the whole sample (see figure 4.3). The distance between the two electrodes is adjustable so that samples of different size can easily be measured. To apply voltages of up to 1100 V a Keithley high-voltage sourcemeter (Model 2410) was used.

4.4   Experimental techniques and improvements

Mainly two different SHG techniques have been applied to investigate the symmetry-breaking order parameters and the related domain structure in magnetically induced ferroelectrics.

Order-parameter-related SHG signals are identified and separated by polarization-dependent SHG spectroscopy, performed at different temperatures, i.e. within the various ordered phases of the investigated systems. The spatial degree of freedom of optical SHG is exploited to image the associated domain structures and reveal spatial correlation effects.

For the first time, these methods have been applied and proved feasible regarding the investigation of incommensurate antiferromagnetic order. Most remarkably, optical SHG appears to be explicitly powerful for visualizing the inherent domains of spin-spiral ferroelectrics.

4.4.1   Nonlinear optical spectroscopy

In the previous sections the properties of the nonlinear susceptibility have been discussed in terms of symmetry, without taking into account its microscopic origin. At its roots, the nonlinear response of a system is due to electronic transitions between an initial state $|i\rangle$ and an excited state $|f\rangle$ separated by the energy gap $\Delta E_{fi} = \hbar\omega_{SHG} = 2\hbar\omega$ (see also fig. 4.1). All possible initial and final states are determined by the electronic structure of the investigated crystal, meaning the band structure or the local crystal-field splitting. Therefore, it is not surprising that the nonlinear susceptibility shows a spectral dependence, leading to explicitly strong SHG signals in the case of a resonant excitation [125,131].

Taking MnO as an example, most of the d–d transitions of the Mn^{2+} ions are found between 2 and 4 eV [132]. Like in all oxides, the localized d electrons are strongly influenced by the crystal field provided by the surrounding octahedra of O^{2+} ions, causing the observed crystal-field splitting. The local situation is quite similar for the Mn^{2+} ions in the magnetically induced ferroelectric $MnWO_4$. Hence, only minor deviations in the excitation energies compared to MnO are expected.

Since the $3d$ electrons are responsible for the magnetism in $MnWO_4$ (see also section 5.1.1), the onset of magnetic ordering will affect the electronic structure and the related d–d transitions. In other words, the transitions observed between 2 and 4 eV are expected to be particularly sensitive to nonzero magnetic and/or electric order parameter(s).

This argumentation defines the generic experimental procedure: First of all, polarization-dependent spectra are taken as function of the second harmonic energy $\hbar\omega_{SHG}$ in the different ordered phases of the investigated system, including the high-temperature phase, which shows no electric and magnetic ordering. A comparision of these spectra allows for a separation of SHG contributions into those which are of purely crystallographic origin and others, which couple to different possible order parameters.

The polarization analysis of SHG signals at a certain photon energy is performed by so-called anisotropy measurements. For this purpose, the polarizer and the analyzer in the setup can be rotated separately or simultaneously by 360°. Anisotropy measurements are applied to identify which component(s) χ_{ijk} contribute a detected SHG signal.

SHG contributions of different origin in general have a different polarization dependence, determined by the associated order parameters via equation (4.13). In the case of identical polarization dependencies, characteristic spectral features can be helpful to distinguish and separate the detected SHG signals. Note that the polarization of the emitted SHG light

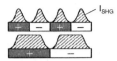

Figure 4.4: Destructive interference of SHG contributions at domain walls in the case of 180° domains [133]. When adjacent domains are attributed to opposite order parameters, indicated by + and −, the phase difference between the related SHG signals is $\Delta\psi = 180°$. As a consequence destructive interference is observed in the vicinity of domain walls marking their path.

is determined by the symmetry properties of the related order parameters, whereas the spectral dependency is caused by the microscopic properties of a system.

4.4.2 Domain imaging

Once the order-parameter-related SHG contributions are identified, spatially resolved measurements of the SHG light allow for imaging of the associated domain structure. Imaging of domains is possible, because the emitted light field carries information about the relative orientation of present order parameters.

This can be understood regarding the components of the nonlinear susceptibility $\hat{\chi}$. Generally, each component $\chi_{ijk...}$ is a complex number[2], consisting of an amplitude $|\chi_{ijk...}|$ and a phase ψ:

$$\chi_{ijk...} = |\chi_{ijk...}|e^{i\psi} , \tag{4.14}$$

while the phase is determined by

$$\tan\psi = \frac{\text{Im } \chi_{ijk...}}{\text{Re } \chi_{ijk...}} . \tag{4.15}$$

From equation (4.7) and equations (4.4) to (4.6) we can further conclude, that the amplitude of an emitted light wave linearly depends on $\chi_{ijk...}$.

Domains of opposite order parameters (180° domains)

The phase sensitivity has an interesting consequence for adjacent domains attributed to antiparallel order parameters $\pm\eta$, called 180° domains (see section 3.4). In this case a cancelation of the SHG signal occurs in the vicinity of domain walls, due to the above mentioned linear coupling between χ_{ijk} and the nonzero order parameter η [133].

Figure 4.4 schematically illustrates the destructive interference of SHG light at domain walls between such 180° domain. Apparently, SHG light of adjacent domains with $+\eta$ and $-\eta$ is out of phase by $\Delta\psi = 180°$. As a consequence, black lines caused by destructive interference are an indicator for the presence of 180° domain walls.

The thickness of the observed black lines in spatially resolved measurements is in general not determined by the thickness of the domain wall, but by the resolution limit of the optical equipment ($\gtrsim 1$ μm).

---

[2]The notation $\chi_{ijk...}$ reflects that equations (4.14) and (4.15) are also valid for tensors of higher rank.

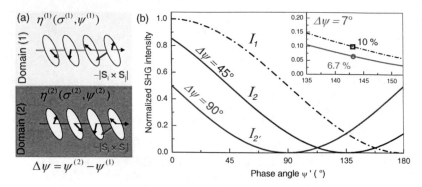

Figure 4.5: Imaging of magnetic translation domains. (a) Schematic illustration of translation domains regarding an elliptical spin-spiral which is for instance observed in the multiferroic phase of $MnWO_4$. Domain walls correspond to a discontinuity $\Delta\psi$ in the phase describing the periodically modulated spin structure. (b) Normalized SHG intensity obtained from domain 1 and domain 2 as function of the tunable phase ψ'. Depending on $\Delta\psi$, different domains exhibit a different level of brightness.

4.4.3   Imaging of translation domains in incommensurate phases

A special type of translation domains emerges in periodically modulated systems with incommensurate long-range order. One example for such translation domains is given in figure 3.9(c), displaying magnetic translation domains in the presence of cycloidal magnetic order. In contrast to the aforementioned 180° domains, the associated order parameters of different translation domains are in general not antiparallel so that no complete cancelation of related SHG appears at the domain walls between the translation domains. Thus, the imaging of translation domains requires a more sophisticated, phase sensitive technique of SHG topography.

Spatially-resolved SHG measurements presented in this thesis prove the general feasibility of phase sensitive SHG topography regarding the imaging of translation domains in incommensurate systems. For the first time, antiferromagnetic translation domains in an incommensurate ordered phase have been imaged (experimental results are shown in section 7.3). In the following, the applied method is discussed in detail.

In section 3.4 it was pointed out that (magnetic) translation domains arise in bulk material, because ordered phases emerge from isolated nucleation centers. When isolated *islands* of a periodically modulated ordered phase grow independently and finally converge, discontinuities in the modulation will appear. This is displayed schematically in figure 4.5(a) showing two adjacent magnetic translation domains. In this example the two domains, denoted as domain (1) and domain (2), are assumed to be associated to an elliptical spin spiral.

As discussed in section 2.2.3 the incommensurate spin spirals of domain (1) and domain (2) can be described by complex magnetic order parameters

$$\eta^{(1)}(\sigma^{(1)}, \psi^{(1)}) \quad \text{and} \quad \eta^{(2)}(\sigma^{(2)}, \psi^{(2)}) , \tag{4.16}$$

respectively, with order parameter components

$$\eta_l^{(1,2)} = \sigma_l^{(1,2)} e^{i\psi_l^{(1,2)}} \ . \tag{4.17}$$

σ denotes the order parameter amplitudes while ψ parametrizes the relative phase of the periodically modulated spin structure ($l = 1, 2$). Thus, the phase difference between the two translation domains in figure 4.5 is given by $\Delta\psi = \psi^{(2)} - \psi^{(1)}$.

We have already seen that the SHG process picks up the symmetry changes imposed by the long-range order by coupling directly to the magnetic order parameters $\eta^{(1,2)}(\sigma^{(1,2)}, \psi^{(1,2)})$. In case of a linear coupling, the relation between the leading SHG contribution from the complex components $\chi_{ijk}^{\text{ED}} = |\chi_{ijk}^{\text{ED}}| e^{i\psi}$ and the order parameter components $\eta_l^{(1,2)}$ is

$$\chi_{ijk}^{\text{ED}} = \chi_{ijkl} \cdot \eta_l^{(1,2)} \ . \tag{4.18}$$

Here, χ_{ijkl} denotes the components of the nonlinear susceptibility above the ordering temperature ($\chi_{ijkl} \equiv \chi_{ijkl}^{(T>T_c)}$). Equation (4.18) indicates that the components χ_{ijk}^{ED} include information about the phase of the magnetic order parameter. However, the phase information is usually lost in SHG experiments, when SHG intensities

$$I \propto |\tilde{\mathbf{S}}|^2 \propto |\chi_{ijk}^{\text{ED}}|^2 \tag{4.19}$$

are measured [134–136][3]. To recover the phase information, a reference signal, described by a plane wave $\tilde{\mathbf{S}}_{\text{ref}}(\psi_{\text{ref}}) = \tilde{\mathbf{S}}_{\text{ref}} e^{i\psi_{\text{ref}}}$, can be superimposed with the actual order-parameter-related SHG signal $\tilde{\mathbf{S}}(\psi) = \tilde{\mathbf{S}} e^{i\psi}$, leading to

$$I \propto |\tilde{\mathbf{S}} + \tilde{\mathbf{S}}_{\text{ref}}|^2 = |\tilde{\mathbf{S}}|^2 + |\tilde{\mathbf{S}}_{\text{ref}}|^2 + 2|\tilde{\mathbf{S}}||\tilde{\mathbf{S}}_{\text{ref}}| \cos(\psi - \psi_{\text{ref}}) \ . \tag{4.20}$$

The interference term in equation (4.20) explicitly depends on the phase of the two interfering light waves and can be exploited to visualize the translation domains as present in spin-spiral ferroelectrics.

As an example we consider two translation domains depicted in figure 4.5(a). Equation (4.20) now indicates that the intensity obtained from each of the two domains depends on the relative phase $\psi^{1,2} - \psi_{\text{ref}}$ in the presence of a SHG reference signal. Hence, a contrast will be observed whenever the magnetic order parameters of domain (1) and domain (2) differ in phase, i.e. $\Delta\psi = \psi^2 - \psi^1 \neq 0$. Moreover, the contrast can be controlled by tuning the amplitude $|\tilde{\mathbf{S}}_{\text{ref}}|$ and the phase ψ_{ref} of the reference signal.

According to equation (4.20), the largest contrast between translation domains is achieved in spatially-resolved SHG measurements when $I_{1,2} \propto |\eta^{(1,2)}(\sigma^{(1,2)}, \psi^{(1,2)})|^2$ and $I_{\text{ref}} \propto |\tilde{\mathbf{S}}_{\text{ref}}|^2$ are of the same magnitude. By defining

$$\sqrt{I_{1,2}} = \sqrt{I_{\text{ref}}} = \frac{1}{2} \ , \tag{4.21}$$

the normalized SHG intensity obtained from domain (1) and domain (2) is

$$I_1 = \frac{1}{2} + \frac{1}{2} \cos(\psi^{(1)} - \psi_{\text{ref}}) \tag{4.22}$$

$$I_2 = \frac{1}{2} + \frac{1}{2} \cos(\underbrace{\psi^{(1)} + \Delta\psi}_{=\psi^{(2)}} - \psi_{\text{ref}}) \ . \tag{4.23}$$

---

[3]Note that the leading electric-dipole contribution to $\tilde{\mathbf{S}}$ is considered.

Apparently, the contrast between the two domains can be controlled by adjusting the relative phase $\psi' := \psi^{(1)} - \psi_{\text{ref}}$.

Figure 4.5(b) shows the normalized SHG intensity obtained from domain (1) and domain (2) as function of ψ', considering selected values for the phase different $\Delta\psi$ of the associated order parameters.

First of all, one can see that for a fixed value of ψ', the contrast between relative domains is determined by the phase difference $\Delta\psi$. Hence, different translation domains exhibit a different level of brightness in spatially-resolved SHG measurements. Most remarkably, the contrast between relative domains can be controlled by tuning ψ' in a way that smallest phase differences are detectable. However, an enhancement of the contrast always coincides with a loss of intensity so that a compromise must be found. The inset in figure 4.5(b) exemplifies the sensitivity towards the phase difference of relative domains. One can see that even phase differences $\Delta\psi$ of only 7° are detectable with a contrast of 1.5 : 1 in spatially-resolved measurements ($\psi' = 143°$), provided that 10% of the initially detected SHG intensity from domain (1) is sufficient to image the domain topology.

In conclusion, it is shown that translation domains in spin-spiral systems can be imaged by SHG topography. They are visualized by superimposing a reference signal with the order-parameter-related SHG contributions which leads to detectable contrasts. By adjusting the contrast, even smallest phase differences between translation domains become visible. The results will be published in the MRS Proceedings (2009) in Meier *et al.*, *Imaging of Hybrid-Multiferroic and Translation Domains in a Spin-Spiral Ferroelectric*.

Chapter 5

Investigated model compounds

For the study of magnetically induced ferroelectrics, $MnWO_4$ and $TbMn_2O_5$ have been chosen as model end compounds with respect to their very different microscopy. $MnWO_4$ is a non-rare-earth compound with Mn^{2+} ($S = 5/2$) in which multiferroicity is driven exclusively by antisymmetric spin-spin interactions whereas $TbMn_2O_5$ is a rare-earth compound with Mn^{4+} ($S = 3/2$) and Mn^{3+} ($S = 2$) in which mainly symmetric spin-spin interactions drive the multiferroic state [96, 103, 114].

In this chapter the relevant properties of both compounds are reviewed. Individual sections are devoted to the complex order parameters of $MnWO_4$ and $TbMn_2O_5$, because understanding their evolution and impact on the point symmetry of the crystals is a prerequisite for the analysis of SHG measurements.

Furthermore, an overview regarding symmetry-allowed SHG contributions in the various ordered phases of the two systems is provided.

5.1 Manganese tungstate – $MnWO_4$

5.1.1 Crystal structure

The mineral huebnerite, or manganese tungstate, $MnWO_4$, belongs to the isomorphous series AWO_4 with $A = $ Mn, Fe, Co, Ni, Zn, Cd and Mg [137]. It crystallizes in a monoclinic way with space group $P2/c1'$ (point group $2/m1'$), and the lattice parameters at room temperature are $a = 4.830$ Å, $b = 5.7603$ Å, and $c = 4.994$ Å. The monoclinic angle $\beta = 91.14°$ in between the a and the c axis is rather close to $90°$. Hence, we will use a Cartesian coordinate system to approximate the monoclinic unit cell, so that $a = x$, $b = y$, and $c \approx z$. The small deviation from the *real* structure is negligable in optical studies, but drastically simplifies the discussion in terms of symmetry.

High-quality single crystals of $MnWO_4$ grown by using the flux technique have been supplied by P. Becker and L. Bohatý[1]. The crystal structure is schematically shown in figure 5.1 [138]. As depicted in figure 5.1(a), zig-zag chains of edge-sharing distorted MnO_6 octahedra running along the z axis alternate with zig-zag chains of edge-sharing distorted WO_6 octahedra. For a clear view, only bonds and octahedra of the MnO_6 chain are indicated. Black solid lines mark the monoclinic unit cell of $MnWO_4$. Furthermore, the top view of the xz plane reveals stacking of alternate Mn and W layers along the x direction (figure 5.1(b)).

---

[1]Institut für Kristallographie, Universität zu Köln

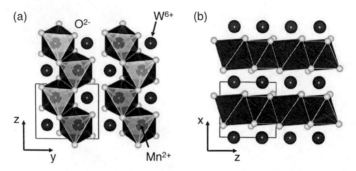

Figure 5.1: Crystal structure and unit cell (black solid lines) of $MnWO_4$. (a) The structure consists of alternating zig-zag chains of edge-sharing MnO_6 octahedra and edge-sharing WO_6 octahedra, running along the z axis. The latter ones are not sketched. (b) The topview of the xz plane reveals an alternate stacking of the Mn and W layers along the x axis.

As already mentioned, the only type of magnetic ion in $MnWO_4$ is Mn^{2+}. Its five $3d$ electrons are in a high-spin configuration, leading to $S = 5/2$. Each of the Mn^{2+} ions has nine different Mn^{2+} neighbors interacting via $90°$ and $180°$ superexchange interactions. This leads to a frustration of the magnetic subsystem [139–142].

5.1.2 Phase diagram

At low temperature, three magnetically ordered phases determine the behavior of $MnWO_4$ in the absence of external fields. The related sequence of phase transitions is sketched in the upper part of figure 5.2[2].

In the AF3 phase below 13.5 K (T_1) incommensurate antiferromagnetic ordering of the Mn^{2+} moments is observed. All spins align in a collinear way along the magnetic easy axis of $MnWO_4$ within the xz plane, inclosing an angle of $\approx 34° - 37°$ with the x axis of the monoclinic crystal (for a simplified illustration see lower part of figure 5.2). Their magnitude is sinusoidally modulated, leading to a two-dimensional incommensurate spin-density wave (SDW), described by the wave vector $\mathbf{k}_{AF3} = (-0.214, \frac{1}{2}, 0.457)$ [141]. The two irrational components of $\mathbf{k}_{AF3}$ imply that the translation symmetry of the magnetically ordered system is violated in x and z direction.

Regarding the symmetry of the AF3 phase, the sinusoidal SDW conserves the center of inversion. Therefore, as we have seen in section 3.3.2, no spontaneous polarization can be induced by magnetism in this phase. This is explicitly expressed by equation (3.14), leading to $\mathbf{P} = 0$ because of the collinear spin arrangement with $\mathbf{S}_i \times \mathbf{S}_j = 0$.

Upon cooling, the system becomes multiferroic across a continuous (second-order) phase transition at $T_2 = 12.7$ K. In this so-called AF2 phase an additional transverse spin component orders and the SDW turns into an elliptical spin spiral without changing the periodicity of the modulation, so that $\mathbf{k}_{AF2} = \mathbf{k}_{AF3} = (-0.214, \frac{1}{2}, 0.457)$. As indicated by

---

[2]Note that the transition temperatures are not denoted uniformly in the literature. Here, for consistency, the temperatures are denoted by T_1 to T_3, starting with the transition which occurs first on lowering the temperature. The three phases are commonly denoted by AF3 to AF1, with AF1 being the magnetic ground state.

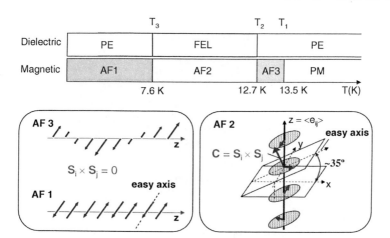

Figure 5.2: Phase transitions and magnetic ordering in MnWO$_4$. Upper part: Sequence of phase transitions and denotation of the three magnetically ordered phases. Lower part: Simplified illustration of the occuring spin arrangements. In the AF3 and AF1 phase spins order collinear along the easy axis, forming an incommensurate two-dimensional sinusoidally modulated spin-density wave and a rather simple antiferromagnetic structure, respectively. In the multiferroic phase an *elliptical* spin spiral violates the spatial inversion symmetry and induces a spontaneous polarization.

figure 5.2, the ellipse associated to the rotation of the spins lies within an easy plane, defined by the easy axis and the principal y axis.

The elliptical spin spiral violates the inversion symmetry of the crystal and induces a spontaneous polarization P_y parallel to the y axis via *antisymmetric* exchange interactions, described by equation (3.14). Remarkably, the vector chirality $\mathbf{C} = \mathbf{S}_i \times \mathbf{S}_j$ associated to the elliptical spin spiral reverses along with the sign of P_y in electric fields [143].

Nevertheless, the magnetically induced polarization corresponds to a secondary order parameter, being merely a consequence of the magnetic ordering that drives the transition at T_2 (see chapter 2.1.4). As common for all known magnetically induced ferroelectrics, the observed spontaneous polarization is rather small with a maximum value of $P_y \approx 5$ nC/cm^2 [96].

In the magnetic ground state (AF1 phase) MnWO$_4$ again displays collinear but now commensurate antiferromagnetic order with spins aligned along the easy axis. The associated wave vector is $\mathbf{k}_{\mathrm{AF1}} = (\pm\frac{1}{4}, \frac{1}{2}, \frac{1}{2})$, indicating simple antiferromagnetic order along y and z, whereas an up-up-down-down configuration of spins is observed along x. Consistent with equation (3.16), the spontaneous polarization vanishes across the discontinuous (first-order) AF2 $\rightarrow$ AF1 transition ($T_3 = 7.6$ K).

So far we only reviewed the zero-field behavior of manganese tungstate. The magnetic-field dependence of the different ordered phases is summarized in figure 5.3.

Magnetic fields applied along the easy axis destabilize the paraelectric AF1 phase, which is already completely suppressed for $H \gtrsim 2$ T in favor of the multiferroic AF2 phase [144]. Furthermore, the stability range of the AF3 phase is shrinking with increasing $H\|$easy axis.

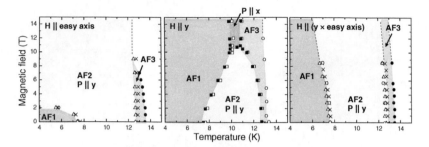

Figure 5.3: Schematic HT-phase diagrams of $MnWO_4$ for magnetic fields applied parallel to the easy axis, the y, axis and a third axis that is perpendicular to the two defined by $y \times$ easy axis. Shown phase boundaries base on measurements published by Arkenbout *et al.* and Taniguchi *et al.* – data points are taken from references [144] and [96].

In contrast, the AF1 and the AF3 phase are stabilized in fields $H||y$. As a consequence, the multiferroic state shrinks with increasing H and, most remarkably, a magnetic field of ≈ 11 T allows to flop the spontaneous polarization **P** from the y to the x direction, reflecting pronounced magnetoelectric interactions in $MnWO_4$.

For completeness, figure 5.3 also displays the phase diagram derived for magnetic fields applied parallel to an axis which is perpendicular to the easy axis and the y axis [144]. In this case, magnetic fields of up to 9 T have only minor influence on the phase boundaries, leading to a slight decrease of the associated transition temperatures for increasing H.

A more extended discussion about field-dependent measurements can be found in the references [96, 144] and [145].

Remark: The sequence of phase transitions in $MnWO_4$ is characteristic for various multiferroic spin-spiral systems as orthorhombic $TbMnO_3$ or $DyMnO_3$, $Ni_3V_2O_8$, or CuO [146]. In each case, an initial incommensurate magnetic order evolves below T_1 and becomes unstable as the temperature is further lowered. Typically, the initial order involves spins oriented along the easy axis of the crystal with a sinusoidally modulated magnitude. Only at an even lower critical temperature T_2 a "fixed length of the spins" becomes energetically favorable and a second transition takes place, at which a transverse component of the spins becomes nonzero (in a semiclassical picture). If this spin arrangement violates the inversion symmetry of the crystal, a magnetically induced polarization can appear.

5.1.3   Symmetry-breaking order parameters

The magnetic and electric order parameters determining the physical properties of $MnWO_4$ at low temperature can be described by the Landau theory for incommensurate phases. In the following, a summary of the most important results is given.

Each of the magnetically ordered phases exhibits a periodically modulated arrangement of spins which can be represented by a general nonzero wave vector $\mathbf{k} = (q_x, \frac{1}{2}, q_z)$. The wave vector $\mathbf{k}$ is incommensurate in the phases AF3 and AF2, and commensurate in the magnetic ground state (AF1). According to section 2.2, the symmetry properties of $\mathbf{k}$ have to be taken into account for a symmetry analysis of the ordered states.

Phase	Point group	Order parameters	Ferroelectric
PARA	$2_y/m_y1'$	$\sigma_1 = 0,\ \sigma_2 = 0$	NO
AF3	$2_y/m_y1'$	$\sigma_1 \neq 0,\ \sigma_2 = 0$	NO
AF2	$2_y1'$	$\sigma_1 \neq 0,\ \sigma_2 \neq 0$	YES
AF1	$2_y/m_y1'$	$\sigma_1' \neq 0,\ \sigma_2' = 0$	NO

Table 5.1: Ordered phases and related point-group symmetries. Nonzero values of the order-parameter amplitudes σ_i indicate which magnetic order parameter(s) is/are nonzero in the various phases, while the last column points out whether the phase is ferroelectric or not.

The wave vector **k** defines different specific irreducible representations $\Gamma_j^{(\mathbf{k})}$ of the high-temperature space group $G_0 = P2/c1'$ and the $\Gamma_j^{(\mathbf{k})}$ are of the same dimension as the symmetry-allowed magnetic order parameters η_j (see section 2.2.2).

Thus, it is sufficient to know the the little group of **k** and the star of **k** for identifying possible magnetic order parameters of MnWO$_4$.

The little group $G(\mathbf{k}) \subset G_0$ of a general $\mathbf{k} = (q_x, \frac{1}{2}, q_z)$ is

$$G(\mathbf{k}) = \left\{ (x,y,z), (x, \bar{y}, z + \frac{1}{2}) \right\} \ .$$

The elements of $G(\mathbf{k})$ are denoted by their effect on the coordinates x, y, z of a general point, while the coordinate system is given by the principle axes of the crystal. This is the standard notation used for the elements of space groups[3] .

By help of the *Bilbao Crystallographic Server* it can easily be calculated that this little group has two one-dimensional small representations $\tau_j(\mathbf{k})$ ($j = 1, 2$) and the star of **k** has the two arms $\pm\mathbf{k}$ ($\ell = 2$) [60]. Hence, according to equation (2.21), two two-dimensional order parameters η_j may arise at low temperature. The components are formed by the complex amplitudes of the magnetic waves, so that the two magnetic order parameters can be represented by:

$$\eta_1 = \left(\begin{array}{c} \sigma_1 e^{i\theta_1} \\ \sigma_1 e^{-i\theta_1} \end{array} \right) \ , \ \eta_2 = \left(\begin{array}{c} \sigma_2 e^{i\theta_2} \\ \sigma_2 e^{-i\theta_2} \end{array} \right) \ . \tag{5.1}$$

All order-parameter invariants and the Landau expansion of the free energy $F(T, \sigma_1, \sigma_2, \theta_1 - \theta_2)$ are given in reference [147]. The relevant results are summarized in table 5.1.

The first magnetic order parameter η_1 arises at T_1 and induces the AF3 phase while the point symmetry $2_y/m_y1'$ of the paramagnetic/-electric phase (PARA) is conserved. Note that the AF3 phase cannot be described by a space group, because the incommensurate magnetic order violates the three-dimensional translation symmetry.

Below T_2 the second magnetic order parameter η_2 emerges and coexists with η_1 in the multiferroic AF2 phase. The order parameter η_1 is considered to be "frozen" at T_2, meaning that the transition is exclusively driven by η_2. The term "frozen" is used to indicate that η_1 shows no critical behavior for $T \leq T_2$. Hence, the order-parameter amplitude σ_1 of η_1 can be assumed to be constant within the AF2 phase (see also figure 5.4).

The influence of the frozen order parameter manifests itself only in the symmetry of the spiral phase. While the free energy in the AF3 phase is independent of the phase θ_1 of

---

[3]Note that $P2/c1'$ is a non-symmorphic space group which is reflected by the elements of $G(\mathbf{k})$, being no "pure" point-symmetry operations.

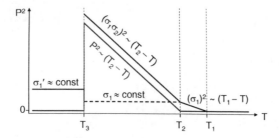

Figure 5.4: Schematic illustration of the temperature dependence of the order parameters in MnWO$_4$: For a simplification of the graph the square of the spontaneous polarization P_y and the magnetic order-parameter amplitudes σ_1 and $\sigma_1\sigma_2$ is shown. As discussed in the text, P_y^2 and σ_i^2 linearly depend on the temperature (Note that σ_1 is in a frozen state below T_2).

η_1, minimization of F in the AF2 phase with respect to θ_1 and θ_2 leads to the equilibrium value $\theta_1 - \theta_2 = (2n+1)\frac{\pi}{2}$ for the relative phase in AF2.

The elliptical spin spiral induced by η_1 and η_2 ($\sigma_1 \neq 0$, $\sigma_2 \neq 0$) violates the center of inversion and reduces the point group of the crystal to $2_y1'$. As a consequence, the magnetic ordering induces a spontaneous polarization

$$P_y = \pm\delta\chi_e\sigma_1\sigma_2 \,, \qquad (5.2)$$

provided that a lowest order coupling term, linear in P_y, η_1, and η_2, is assumed to contribute to the free energy [147]. Here, δ denotes the coupling constant.

As mentioned before, σ_1 is assumed to be constant in the AF2 phase. It can be either positive or negative, but sign and magnitude are fixed. Thus, any reversal of P_y is exclusively linked to the order parameter η_2 which is the only *active* order parameter and the sign of P_y reverses along with the sign of the order-parameter amplitude σ_2. Hence, we can conclude that η_2 is uniquely related to the sense of the elliptical spin spiral of MnWO$_4$, i.e. the vector chirality $\mathbf{C} := \mathbf{S}_i \times \mathbf{S}_j$ (see section 5.1.2).

Regarding the scaling behavior of the ferroelectric polarization P_y, equation (5.2) predicts a linear dependence on the primary magnetic order parameter, i.e. its amplitude $\sigma_2 \propto (T_2 - T)^{\frac{1}{2}}$. As a consequence, the magnetically induced polarization scales like $P_y \propto (T_2 - T)^{\frac{1}{2}}$ which is in agreement the experimental results of reference [144].

Note that the observed scaling behavior of P_y is quite remarkable, since the ferroelectric polarization is a secondary order parameter that only arises in cause of a coupling to the primary magnetic order parameter. Usually such secondary order parameters scale linear with the temperature as it was shown in section 2.1.4. In the present case the improper ferroelectric polarization behaves exactly as a primary order parameter.

This scaling behavior of P_y clearly separates the multiferroic spin-spiral system MnWO$_4$ from conventional improper ferroelectrics and classifies the system as a so-called *pseudo-proper* ferroelectric in the multiferroic AF2 phase.

The spontaneous polarization quenches at the first-order AF2 $\rightarrow$ AF1 transition, indicating a decoupling of the two magnetic order parameters at T_3. Only one magnetic order parameter, either η_1 or η_2, is nonzero in the AF1 phase. However, due to the first-order nature of the transition one cannot say whether η_1 or η_2 is the relevant parameter. Here, we arbitrarily denote the order parameter of the AF1 phase by

$$\eta'_1 = \begin{pmatrix} \sigma'_1 e^{i\theta_1} \\ \sigma'_1 e^{-i\theta_1} \end{pmatrix} ,$$

with $\theta_1 = \frac{\pi}{4}$. The prime indicates that η'_1 differs from η_1, now describing a commensurate magnetic phase.

Hence, the ground state is described by a standard magnetic space group C_a2/c [141,147]. The associated point group is $2_y/m_y1'$. Remarkably, even the time-inversion symmetry is conserved, because the inversion of all spins is equivalent to a simple translation along y by one lattice constant, due to the antiferromagnetic ordering in this direction ($k_y = \frac{1}{2}$). Figure 5.4 finally summarizes schematically the reported scaling behavior of the magnetic order parameters and the induced ferroelectric polarization. For simplification, the square of P_y is sketched, because P_y^2 exhibits a linear dependence on the temperature, just as the square of the order-parameter amplitudes σ_1^2 and σ_2^2.

5.2 SHG contributions in MnWO$_4$

The knowledge about the point-group symmetry of the different ordered phases (table 5.1) allows to deduce the symmetry-allowed SHG contributions in MnWO$_4$.

The high-temperature paraelectric/-magnetic, the AF3, and the AF1 phase are centrosymmetric with point symmetry $2_y/m_y1'$. Thus, no SHG signals of electric-dipole type are symmetry-allowed in these phases [47]. Nevertheless, higher-order SHG contributions of magnetic-dipole (MD) or electric-quadrupole (EQ) type can emerge[4].

However, EQ contributions for point symmetry $2_y/m_y1'$ have the same polarization dependence as the MD contributions and therefore cannot be distinguished. Thus, we arbitrarily denote the EQ contributions in the same way as the MD contributions ($\hat{\chi}^{MD}$), keeping in mind that detected signals can either be of MD or EQ type. All symmetry-allowed tensor components χ_{ijk}^{MD} of the nonlinear susceptibility for point symmetry $2_y/m_y1'$ are listed in the upper part of table 5.2. They are evaluated on the basis of equations (4.4) to (4.7), using general tensor relations given in reference [47].

In the multiferroic AF2 phase the inversion symmetry is violated and the system has point symmetry $2_y1'$. As a consequence, additional SHG contributions of ED type become symmetry-allowed in between T_2 and T_3 (see lower part of table 5.2).

The overview of tensor components in table 5.2 yield different important information regarding investigations by SHG. Generally, no SHG signals are expected for light fields incident parallel to the y axis of MnWO$_4$.

In contrast, contributions of MD type are symmetry-allowed at all temperatures for $k||x$ and $k||z$. Such MD contributions may already be present in the paraelectric/-magnetic phase for $T > T_1$, constituting a crystallographic background which possibly superimposes with additional order-parameter-related SHG signals below T_1. Nevertheless, the contributions of MD type can be sensitive to the different phase transitions in MnWO$_4$, too, meaning that related SHG signals either emerge at T_i or change in intensity.

Most interesting, SHG contributions of ED type are restricted to the multiferroic AF2 phase. Hence, one can conclude that SHG contributions related to components χ_{ijk}^{ED} provide

---

[4]A more detailed discussion about how to derive the allowed tensor components in MnWO$_4$ can be found in reference [148].

Point group	Incident light field	Symmetry-allowed tensor components (MD)	Symmetry-allowed tensor components (ED)
$2_y/m_y1'$	$k\|x$	$\chi_{yyy}^{\mathrm{MD}}, \chi_{yzz}^{\mathrm{MD}}, \chi_{zyz}^{\mathrm{MD}}$	—
	$k\|y$	—	—
	$k\|z$	$\chi_{yyy}^{\mathrm{MD}}, \chi_{yxx}^{\mathrm{MD}}, \chi_{xyx}^{\mathrm{MD}}$	—
	$k \nparallel x,y,z$	$\chi_{xyz}^{\mathrm{MD}}, \chi_{yxz}^{\mathrm{MD}}, \chi_{zxy}^{\mathrm{MD}}$	—
$2_y1'$	$k\|x$	$\chi_{yyy}^{\mathrm{MD}}, \chi_{yzz}^{\mathrm{MD}}, \chi_{zyz}^{\mathrm{MD}}$	$\chi_{yyy}^{\mathrm{ED}}, \chi_{yzz}^{\mathrm{ED}}, \chi_{zyz}^{\mathrm{ED}}$
	$k\|y$	—	—
	$k\|z$	$\chi_{yyy}^{\mathrm{MD}}, \chi_{yxx}^{\mathrm{MD}}, \chi_{xyx}^{\mathrm{MD}}$	$\chi_{yyy}^{\mathrm{ED}}, \chi_{yxx}^{\mathrm{ED}}, \chi_{xyx}^{\mathrm{ED}}$
	$k \nparallel x,y,z$	$\chi_{xyz}^{\mathrm{MD}}, \chi_{yxz}^{\mathrm{MD}}, \chi_{zxy}^{\mathrm{MD}}$	$\chi_{xyz}^{\mathrm{ED}}, \chi_{yxz}^{\mathrm{ED}}, \chi_{zxy}^{\mathrm{ED}}$

Table 5.2: Symmetry-allowed components $\chi_{ijk}^{\mathrm{MD,ED}}$ of the nonlinear susceptibility for light propagating along the x, y, and z direction of $MnWO_4$. Whenever k is not parallel to one of the principle axes SHG contributions with $i \neq j \neq k$ may admix.

access to the active magnetic order parameter η_2 and, because of equation (5.2), to the ferroelectric ordering P_y.

Note that the point symmetry $2_y1'$ of the multiferroic phase is deduced by assuming two coexisting magnetic order parameters η_1 and η_2. However, considering instead only the nonzero spontaneous polarization P_y also leads to point symmetry $2_y1'$ for the AF2 phase. This implies that the secondary ferroelectric order parameter allows for exactly the same SHG contributions as the coexisting magnetic order parameters, even without taking into account equation (5.2).

The correlation between the imposed symmetry reduction by the magnetic/electric order parameters and nonzero components of the nonlinear susceptibility tensor are the basis for the later analysis of the experimental results.

5.3   Terbium manganite – $TbMn_2O_5$

5.3.1   Crystal structure

The crystal structure of $TbMn_2O_5$ is based on an orthorhombic unit cell of space group $Pbam1'$ (point group $mmm1'$) with lattice constants $a = 7.325$ Å, $b = 8.501$ Å, and $c = 5.666$ Å [149][5]. For our study, flux-grown single crystals of $TbMn_2O_5$ have been provided by V. A. Sanina and R. Pisarev[6]. As a member of the RMn_2O_5 series (R=Y, rare-earth, Bi and La), it consists of edge-sharing $Mn^{4+}O_6$ octahedra along the z axis, and pairs of $Mn^{3+}O_5$ pyramids linking with two $Mn^{4+}O_6$ chains, schematically shown in figure 5.5(a) [150,151].

The structure contains three types of magnetic ions (Mn^{4+} ($3d^3$), Mn^{3+} ($3d^4$), and Tb^{3+} ($4f^8$)) being responsible for the complex magnetic ordering observed at low temperature. There are five distinct magnetic interactions (J_1 to J_5) between neighboring Mn spins, mediated through a direct Mn^{4+}–Mn^{4+} path and three paths with intermediary oxygen

---

[5]In the later discussion of the experimental results, a Cartesian coordinate system with $a = x$, $b = y$, and $c = z$ will be used to represent the crystallographic axes.

[6]Ioffe Physical Technical Institute, Russian Academy of Science, St. Petersburg

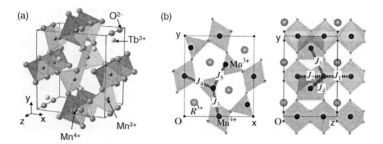

Figure 5.5: Crystal structure and magnetic exchange interactions J_i of TbMn$_2$O$_5$. (a) The orthorhombic structure is composed of edge-sharing MnO$_6$ octahedra along the z axis, linked by pairs of MnO$_5$ pyramids [116]. (b) The display projections onto the xy and the yz plane reveal five different exchange interactions J_1 to J_5 between Mn ions, causing a pronounced frustration of the magnetic sublattice [152].

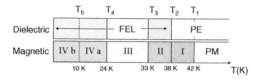

Figure 5.6: Illustration of the sequence of magnetic and ferroelectric phase transitions in TbMn$_2$O$_5$. Four different incommensurate (I, II, and IV) and commensurate (III) magnetically ordered phases determine the behavior at low temperature. Besides phase I, all phases show multiferroic properties in zero field. The Tb sublattice orders at T_5 so that phase IV can further be subdevided with respect to disordered (IVa) and ordered (IVb) Tb spins.

(Mn^{3+}–O–Mn^{4+}, Mn^{3+}–O–Mn^{3+}, and Mn^{4+}–O–Mn^{4+}), depicted in figure 5.5(b) [153]. When the magnetic moments of the Tb ions contribute, additional exchange interactions have to be included.

The manifold competing magnetic interactions cause pronounced frustration that leads to various antiferromagnetic phases in the low-temperature regime. The different phases are discussed in the following section.

5.3.2 Phase diagram

The sequence of phase transitions in TbMn$_2$O$_5$ is sketched in figure 5.6. Overall, five phase transitions are observed upon cooling, occuring successively at $T_1 = 42$ K, $T_2 = 38$ K, $T_3 = 33$ K, $T_4 = 24$ K, and $T_5 = 10$ K [9,101,154].

First, we consider the series of phase transitions in terms of the associated wave vectors. The magnetic order of the phases I and II is described by a temperature-dependent incommensurate wave vector $\mathbf{k}_1 = (\frac{1}{2}, 0, q_z)$, with q_z decreasing from 0.276 to $\frac{1}{4}$ on lowering the temperature[7].

---

[7]Here, the wave vector subscript $i = 1$ indicates the associated transition temperature $T_{i=1}$ at which the related modulation occurs.

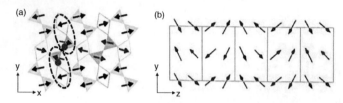

Figure 5.7: Ordering of the Mn moments in the commensurate phase II of RMn$_2$O$_5$ [40].
(a) Strong antiferromagnetic nearest-neighbor interactions within the xy plane lead to a mainly
antiparallel spin arrangement. (b) For some compounds of the RMn$_2$O$_5$ series a weak antisymmetric exchange is assumed to induce periodically modulated cycloidal magnetic order
along z.

At T_3 the wave vector component q_z *locks in*, meaning that phase III is related to a
constant commensurate wave vector $\mathbf{k}_3 = (\frac{1}{2}, 0, \frac{1}{4})$. This commensurate phase III is stable
between T_3 and T_4, whereas the magnetic ground state of the Mn sublattice ($T < T_4$)
is again incommensurate with $\mathbf{k}_4 = (0.48, 0, 0.32)$. Finally, the Tb 4f momemts order at
$T_5 \approx 10$ K.

As indicated by color and hatching in the upper part of figure 5.6, the magnetic order
in TbMn$_2$O$_5$ is accompanied by ferroelectric ordering ($\mathbf{P}||y$) in the phases II to IV. The
spontaneous polarization is magnetically induced (improper) and reaches a maximum value
of $P_y = 40$ nC/cm^2 [9, 101].

To emphasize that mainly symmetric exchange interactions are responsible for the multiferroicity of TbMn$_2$O$_5$, we now discuss the specific magnetic structures associated to the
magnetic wave vectors $\mathbf{k}_i$ ($i = 1, 3, 4$) [101, 108, 155].

Excluding subtle variations, the magnetic structures within the RMn$_2$O$_5$ series are quite
similar. They can be understood in terms of magnetic xy planes coupling along the z
direction. The interactions in the xy plane are expected to be uniformly antiferromagnetic
with particularly strong coupling along the x axis. Therefore, antiferromagnetic zig-zag
chains along this direction are common for all magnetically ordered phases. The typical
arrangement of spins in the commensurate phase III is sketched in figure 5.7(a). The spin
structure in phase IV is quite similar, despite the fact that the antiferromagnetic structure
exhibits an incommensurate period along x.

At present it is not known what determines the z axis stacking, but along the z direction quasi-parallel and quasi-antiparallel stacking is observed. A possible origin of the
complex incommensurate stacking could be either a competition of nearest- and next-
nearest-neighbor interactions or an antisymmetric magnetic exchange. For some of the
RMn$_2$O$_5$ compounds also a small z component of the spins is reported for the commensurate phase III, leading to cycloidal magnetic ordering along z (see figure 5.7(b)).

According to Radaelli et al., the magnetically induced ferroelectricity in TbMn$_2$O$_5$ is predominantly caused by symmetric exchange interactions [156]. Most important are the
in-plane components of the magnetic structure in the xy planes, inducing a polar ionic
displacement via exchange striction as discussed in section 3.3.2.

However, noncollinear spins always allow for nonzero spin supercurrents so that the spontaneous polarization might also be of purely electronic origin (see section 3.3.4) or a combination of both. Since no ionic displacements have be measured yet, there is a controversial
dispute concerning the microscopy of magnetically induced ferroelectricity in this com-

Phase	Point group	Order parameter	Phase angle	Ferroelectric
PARA	$mmm1'$	$\sigma_1 = 0$, $\sigma_2 = 0$	–	NO
I	$mmm1'$	$\sigma_1 \neq 0$, $\sigma_2 = 0$	$\theta_1 \neq 0$	NO
II	$m2m$	$\sigma_1 \neq \sigma_2 \neq 0$	$\theta_1 - \theta_2 = (2n+1)\frac{\pi}{2}$	YES
III	$m2m$	$\sigma_1 \neq \sigma_2 \neq 0$	$\theta_1 = \pm\theta_2 = n\frac{\pi}{4}$	YES
IV	$m2m$	$\sigma_1 = \sigma_2 \neq 0$	$\theta_1 - \psi_1 = (2n+1)\frac{\pi}{2}$	YES
		$\rho_1 = \rho_2 \neq 0$	$\theta_2 - \psi_2 = n\pi$	

Table 5.3: Magnetically ordered phases I to IV in TbMn$_2$O$_5$ and associated point-group symmetries. The values σ_i and δ_i indicate which order-parameter components contribute in the different phases. The θ_i and ψ_i denote the equilibrium value of the phase angle related to each order parameter. The last column identifies the ferroelectric. i.e. multiferroic phases.

pound [152].

5.3.3 Symmetry-breaking order parameters

In its high-temperature paraelectric/-magnetic phase, TbMn$_2$O$_5$ has space-group symmetry $G_0 = Pbam1'$. Here, the little group $G(\mathbf{k}) \subset G_0$ of a general wave vector $\mathbf{k} = (\frac{1}{2}, 0, q_z)$ is given by

$$G(\mathbf{k}) = \left\{ (x, y, z), (\bar{x}, \bar{y}, z), (x - \frac{1}{2}, \bar{y} - \frac{1}{2}, z), (\bar{x} - \frac{1}{2}, y - \frac{1}{2}, z) \right\} .$$

The same analysis as discussed in section 2.2 and 5.1.3 reveals that only one two-dimensional small representation $\tau_1(\mathbf{k})$ of $G(\mathbf{k})$ exists, while the star of $\mathbf{k}$ consists of two vectors [60]. Therefore, according to equation (2.21), there is only one four-dimensional irreducible representation $\Gamma_1^{(\mathbf{k})}$ of $Pbam1'$.

Note that the analysis of $\mathbf{k} = (\frac{1}{2}, 0, q_z)$ is representative for $\mathbf{k}_1$ and $\mathbf{k}_3$. The corresponding magnetic order parameter η_1 has four components which are given by the complex amplitudes of the magnetic waves:

$$\eta_1 = \begin{pmatrix} \sigma_1 e^{i\theta_1} \\ \sigma_1 e^{-i\theta_1} \\ \sigma_2 e^{i\theta_2} \\ \sigma_2 e^{-i\theta_2} \end{pmatrix} . \tag{5.3}$$

This four-dimensional order parameter describes the three magnetically ordered phases I, II, and III. Order-parameter invariants and the minimization of the free-energy expansion $F(\sigma_1, \sigma_2, \theta_1 - \theta_2)$ are the subject of reference [92]. The main results regarding the equilibrium values of σ_i and θ_i, as well as the point symmetry of the different stable states are summarized in table 5.3.

Phase I displays point symmetry $mmm1'$ and minimization of $F(\sigma_1, \sigma_2, \theta_1 - \theta_2)$ with respect to the magnetic order parameters reveals $\sigma_1 \neq 0$ and $\sigma_2 = 0$ in phase I. Since only one order-parameter amplitude is nonzero, the inversion symmetry is conserved (see also section 5.1.3).

In phase II, both order-parameter components are nonzero with a fixed relation between their phase angles θ_1 and θ_2. The equilibrium value of the phase difference is $\theta_1 - \theta_2 = (2n+1)\frac{\pi}{2}$. As a consequence of the nonzero phase difference, inversion symmetry is broken and a ferroelectric polarization P_y shows up in phase II.

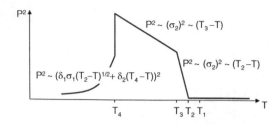

Figure 5.8: Schematic illustration of the temperature dependence of the spontaneous polarization P_y and the magnetic order-parameter amplitude in TbMn$_2$O$_5$. For a simplification of the graph, the square of the aforementioned values is shown. An explanation of the different formula is given in the text.

Note that the situation is qualitatively different compared to MnWO$_4$. In MnWO$_4$ the coexistence of two magnetic order parameters violates the inversion symmetry of the system, whereas independent components of only *one* magnetic order parameter cause the symmetry reduction in the present case. Nevertheless, the relation between P_y and the order-parameter amplitudes σ_i is quite similar [92]

$$P_y = \pm \delta_1 \chi_e \sigma_1 \sigma_2 \, , \tag{5.4}$$

with δ_1 being the coupling constant between the primary magnetic and the secondary ferroelectric order parameter.

The order-parameter components including σ_1 induce the phase transition at T_1. However, for $T \leq T_2$ they are frozen so that the phase transition at T_2 is exclusively driven by the order-parameter components related to σ_2. The latter ones are said to be the *active* order-parameter components for $T \leq T_2$. Since σ_1 is frozen in phase II, it can assumed to be constant.

This leads to an effective linear coupling of P_y and σ_2 simulating proper ferroelectric behavior, meaning $P_y \propto (T_2 - T)^{\frac{1}{2}}$. Nevertheless, P_y is related to a secondary order parameter and emerges merely as a consequence of the magnetic ordering. Hence, TbMn$_2$O$_5$ acts as a pseudo-proper ferroelectric in phase II.

At T_3 the magnetic order becomes commensurate ($\mathbf{k}_3 = (\frac{1}{2}, 0, \frac{1}{4})$), modifying the symmetry of η_1. The main difference between phase II and III is that the translation symmetry of phase III gives rise to additional invariants leading to $\theta_1 = \pm\theta_2 = n\frac{\pi}{4}$, whereas the point symmetry remains $m2m$.

Moreover, the II $\rightarrow$ III transition is a continuous (second order) transition, not changing the general scaling behavior of P_y. Solely the coupling constant δ_1 changes at T_3, causing a slight change in the slope of the temperature dependence of P_y. This is schematically shown in figure 5.8. Note that for simplification of the graph the square of the polarization P_y^2 is sketched, depending linearly on the temperature.

The scaling behavior of the spontaneous polarization drastically changes at T_4, where the wave vector becomes incommensurate again. Since the wave vector $\mathbf{k}_4 = (0.48, 0, 0.32)$ has two irrational components, its little group differs from the little group of $\mathbf{k}_1$ and $\mathbf{k}_3$. It consists of only to symmetry operations:

$$G(\mathbf{k}_4) = \left\{ (x, y, z), (x - \frac{1}{2}, \bar{y} - \frac{1}{2}, z) \right\} \, .$$

$G(\mathbf{k}_4)$ has two one-dimensional irreducible representations $\tau_{1,2}(\mathbf{k}_4)$, whereas the star of $\mathbf{k}_4$ contains four arms ($\Rightarrow \dim(\Gamma_{1,2}(\mathbf{k}_4)) = 4$). This means that two four-dimensional irreducible representations of $Pbam1'$ exist in phase IV, each being associated to one four-dimensional order parameter.

Hence, two order parameters possibly arise for $T < T_4$, denoted by

$$\eta_1 = \begin{pmatrix} \sigma_1 e^{i\theta_1} \\ \sigma_1 e^{-i\theta_1} \\ \sigma_2 e^{i\theta_2} \\ \sigma_2 e^{-i\theta_2} \end{pmatrix} , \quad \eta_2 = \begin{pmatrix} \rho_1 e^{i\psi_1} \\ \rho_1 e^{-i\psi_1} \\ \rho_2 e^{i\psi_2} \\ \rho_2 e^{-i\psi_2} \end{pmatrix} . \tag{5.5}$$

The minimization of the free energy with resepct to η_1 shows that in phase IV only non-polar structures are stable which is contradictory to the experimental results (the same is valid regarding η_2 instead of η_1) [92].

Only by assuming coexisting order parameters η_1 and η_2, the ferroelectric behavior of phase IV can be reproduced. It is reasonable to assume that η_1 is the same order parameter being already present above T_4, whereas the second order parameter η_2 arises at T_4. The coexistence of two magnetic order parameters is further supported by the experimental observation of multiple magnetic orders, being present in the ferroelectric incommensurate phase IV [157]. All equilibrium values ρ_i, σ_i, θ_i, and ψ_i of phase IV are given in table 5.3 [92].

However, the important information is that the incommensurate phase IV has still point-group symmetry $m2m$, although two magnetic order parameters η_1 and η_2 now participate in the symmetry-breaking mechanism. The equilibrium value of P_y below T_4 given by

$$P_y = \pm\delta_2\chi_e\sigma_1\rho_1 , \tag{5.6}$$

indicating the typical linear scaling behavior of a secondary order parameter of P_y in phase IV (see section 2.1.4). This characteristic scaling behavior of improper ferroelectrics follows from the fact that both order parameters are active for $T \leq T_4$ and each order-parameter amplitude, σ_1 and ρ_1, scales like $\propto (T_4 - T)^{\frac{1}{2}}$. *Active* here means that on the one hand the equilibrium values σ_i and θ_i associated to η_1 change at T_4. On the other hand η_2 emerges at T_4 and continuously increases with decreasing temperature.

The superposition of the distinct contributions to the spontaneous polarization given by equations (5.4) and (5.6) reproduces the experimentally observed temperature dependence $P_y(T)$ [9]:

$$P_y(T) = \pm\chi_e(\underbrace{\delta_1\sigma_1(T_2 - T)^{\frac{1}{2}}}_{>0} + \underbrace{\delta_2(T_4 - T)}_{<0}) , \tag{5.7}$$

with $\delta_1 > 0$, and $\delta_2 < 0$. Equation (5.7) furthermore emphasizes that the net polarization is composed of two opposite components which appear at T_2 and T_4, respectively. The evolution of a second ferroelectric contribution, oriented opposite to the first one, explains the jump-like decrease of P_y at T_4.

Most remarkably, the Landau theory of incommensurate phase transitions is the only theoretical framework so far, that allows a consistent description of all consecutive phase transitions in TbMn$_2$O$_5$. The expected temperature dependence of P_y in the complete low-temperature regime is schematically depicted in figure 5.8.

Point group	Incident light field	Symmetry-allowed tensor components (ED)
$m2m$	$k\|\|x$	χ^{ED}_{yyy}, χ^{ED}_{yzz}, χ^{ED}_{zyz}
	$k\|\|y$	—
	$k\|\|z$	χ^{ED}_{yyy}, χ^{ED}_{yxx}, χ^{ED}_{xyx}

Table 5.4: The coexisting magnetic order parameters and the induced ferroelectric order in TbMn$_2$O$_5$ reduce the point symmetry to $m2m$ considering the multiferroic phases II to IV. The point group $m2m$ allows for the listed tensor components to contribute to SHG of ED type. The table indicates that order-parameter-related SHG is expected for incident light waves propagating along the x or z axis of the crystal. In contrast, no SHG is of ED type is symmetry-allowed for $k\|\|y$.

5.4   SHG contributions in TbMn$_2$O$_5$

The point-group symmetry of TbMn$_2$O$_5$ in phases I to IV is given in table 5.3. One can see that the system is centrosymmetric in the paraelectric/-magnetic phase and in the antiferromagnetic phase I, implying that no SHG signals of ED type are symmetry-allowed above T_2 (point group $mmm1'$).

In contrast, the spatial inversion symmetry is violated in the multiferroic phases below T_2 (point group $m2m$). Since the multiferroic phases II to IV lack inversion symmetry, SHG contributions of ED type may emerge below T_2. The symmetry-allowed components χ^{ED}_{ijk} of the nonlinear susceptibility are summarized in table 5.4, considering light fields incident parallel to the principle axes of TbMn$_2$O$_5$. They are evaluated on the basis of equation (4.4) and general tensor relations [47].

Table 5.4 reveals that SHG contributions of ED type are only expected for light fields incident parallel to the x or the z axis of the crystal, while being absent for $k\|\|y$.

Furthermore, one can conclude that all signals of ED type detected with $k\|\|x$ or $k\|\|z$ couple to the magnetic order parameter η_2 which arises at T_2 and, because of equations (5.4) and (5.6), to P_y. Thus, nonzero components χ^{ED}_{ijk} allow to probe the magnetic and/or the electric subsystem of TbMn$_2$O$_5$.

Chapter 6

Magnetically-induced ferroelectricity in $MnWO_4$ and $TbMn_2O_5$

This chapter is devoted to the analysis of the complex magnetic and electric order of magnetically-induced ferroelectrics. We have applied polarization-dependent SHG spectroscopy to investigate the correlation between the inherent magnetic and electric order parameters, using $MnWO_4$ and $TbMn_2O_5$ as model compounds. Different SHG spectra taken in their multiferroic phases are presented and discussed with respect to the symmetry-breaking magnetic and electric order parameters. In addition, temperature-dependent measurements indicate the magnetic and electric phase transitions and reveal the specific temperature-dependence of the order parameter that drives the transition.
Parts of the results are published in Meier et al., PRL **102**, 107202 (2009) and Lottermoser et al. PRB **80**, 1001001(R) (2009).

6.1 Nonlinear spectroscopy of $MnWO_4$

6.1.1 Linear absorption and crystallographic SHG contributions

With respect to the later analysis of order-parameter-related SHG contributions, it is helpful to briefly discuss the linear absorption spectrum and crystallographic SHG in $MnWO_4$. SHG of crystallographic origin is present at all temperatures and thus will superimpose with the relevant order-parameter-related SHG contributions at low temperature.
Figure 6.1(a) shows selected polarization-dependent linear absorption spectra of $MnWO_4$, measured at 295 K and 30 K by using a standard Fourier spectrometer (Bruker IFS). Their structure provides information about electronic transitions and, therefore, indicates at which photon energies strong SHG contributions can be expected due to resonant excitations (see section 4.4.1). The spectra are gained with x-polarized light propagating along the y axis of the crystal. Only marginal variations are obtained with y- or z-polarized light. A comparison of the two curves in figure 6.1(a) reveals that the optical absorption slightly decreases with decreasing temperature, whereas only minor changes are observed regarding the spectral dependence. Both linear absorption spectra show a steep increase of the optical absorption above 2.7 eV, corresponding to the lowest O–Mn charge-transfer transition [158]. Below the charge-transfer transition, two broad peaks are obtained at low temperature (30 K), marked by the black arrows in figure 6.1(a). The two peaks at photon energies of 2.15 eV and 2.65 eV can be assigned to intra-atomic d–d transitions of the Mn^{2+} ions, denoted by $^6A_{1g} \rightarrow {}^4T_{1g}$ and $^6A_{1g} \rightarrow {}^4T_{2g}$, respectively [132, 159, 160].

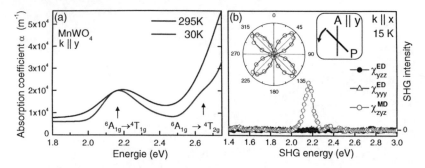

Figure 6.1: (a) Linear absorption spectra of MnWO$_4$, measured with x-polarized light propagating along the y axis of the crystal. At low temperature (30 K) two intra-atomic transitions are identified, whereas the steep increase of the linear absorption above 2.7 eV corresponds to the lowest O–Mn charge-transfer transition. (b) SHG spectra obtained in the paramagnetic/-electric phase of MnWO$_4$. SHG from χ^{MD}_{zyz} is observed at 2.15 eV, being of purely crystallographic origin. The inset displays the anisotropy of the signal, gained by rotating the polarizer (P) while only y-polarized SHG can pass the analyzer (A).

Thus, according to figure 6.1(a), an explicitly strong nonlinear response may occur at SHG energies of 2.15 eV, 2.65 eV and $\approx$ 2.7 eV.

However, it is not compulsory that a nonlinear response couples to the magnetic or electric subsystem. The symmetry analysis in section 5.2 revealed that crystallographic SHG contributions of MD type are symmetry-allowed at all temperatures. At low temperature, these contributions potentially superimpose with the relevant order-parameter-related SHG and, therefore, have to be identified. Fortunately, SHG of crystallographic origin can easily be identified by measuring polarization-dependent SHG spectra in the paramagnetic/-electric phase.

As an example we consider polarization-dependent SHG spectra obtained with $k\|x$ in the paramagnetic/-electric phase of MnWO$_4$, presented in figure 6.1(b). One can see that for light fields incident along the x axis of the crystal, SHG light from χ^{MD}_{zyz} is observed at 2.15 eV, corresponding to the aforementioned lower intra-atomic d–d transition. The assignment of the obtained peak to χ^{MD}_{zyz} is consistent with the symmetry analysis (see table 5.2) and confirmed by the measured rotational anisotropy which is shown in the inset of figure 6.1(b). The rotational anisotropy indicates that y-polarized SHG is observed for incident light fields with nonzero components E_y and E_z[1]. In contrast, no SHG is detected for solely y- or z-polarized light in a configuration with $k\|x$ at 15 K.

An overview of all detected crystallographic SHG contributions in MnWO$_4$ can be found in reference [148]. For the following discussion it is sufficient to keep in mind that these contributions are present at all temperatures and may interfere with order-parameter-related SHG signals in the magnetic and multiferroic phases of MnWO$_4$.

---

[1]Note that the first index of χ^{MD}_{ijk} does not coincide with the polarization direction of the SHG signal in the context of MD contributions. This is expressed by equation (4.7).

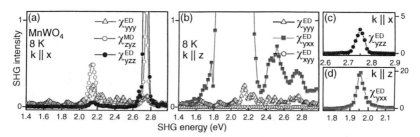

Figure 6.2: Polarization-dependent SHG spectra, obtained in the multiferroic phase of MnWO$_4$. (a) SHG spectra measured with $k\|x$. Two independent contributions from χ^{ED}_{yyy} and χ^{ED}_{yzz} are observed that couple to the multiferroic order. (b) SHG spectra obtained with $k\|z$. Besides SHG from χ^{ED}_{yyy}, a pronounced contribution from χ^{ED}_{yxx} is observed in the multiferroic phase. (c, d) Comparison of the different contributions. SHG from χ^{ED}_{yzz} (χ^{ED}_{yxx}) exceeds SHG from χ^{ED}_{yyy} by a factor of 50 (200).

6.1.2 Multiferroic SHG contributions

The symmetry analysis in section 5.2 indicates that SHG contributions of ED type are symmetry-allowed in the multiferroic phase of MnWO$_4$ only. According to table 5.2, they are accessible with light fields propagating along the x or z axis of the crystal. In addition, SHG signals of MD type that couple to the magnetic or electric subsystem may emerge at T_2, or, if already present above T_2, change in intensity.

To identify the components χ^{ED}_{ijk} and χ^{MD}_{ijk} that are sensitive to the magnetic order parameter η_2 and, via equation (5.2), to the ferroelectric ordering P_y, we measured polarization-dependent SHG spectra in the multiferroic phase at 8 K.

Figure 6.2(a) shows SHG spectra gained with $k\|x$. A pronounced spectral dependence is observed regarding the tensor components χ^{ED}_{yzz}, χ^{ED}_{yyy}, and χ^{MD}_{zyz} that, consistently with table 5.2, contribute to SHG in the AF2 phase. By a comparison with figure 6.1(a), the obtained peaks at 2.15 eV and 2.75 eV can be assigned to the intra-atomic $^6A_{1g} \rightarrow {}^4T_{1g}$ excitation and the lowest O–Mn charge-transfer transition, respectively.

The most prominent peak in figure 6.2(a) is observed at 2.75 eV and corresponds to SHG from χ^{ED}_{yzz} (see also figure 6.2(c)). In comparison, SHG from χ^{ED}_{yyy} is rather weak, exhibiting a maximum in the intensity at 2.15 eV.

Note that the component χ^{MD}_{zyz} in figure 6.2(a) is of purely crystallographic origin and does not couple to the magnetic or electric subsystem. Although the SHG intensity from χ^{MD}_{zyz} detected at 2.15 eV slightly increases in the multiferroic phase compared to the paramagnetic/-electric phase (see figure 6.1(b)), the increase has to be attributed to admixing of SHG from χ^{ED}_{yyy}: While probing SHG from χ^{MD}_{zyz} the polarization of the light wave and the z axis (and y axis) of the crystal enclose an angle of 45°. Thus, the incident light field has nonzero components E_z and E_y, leading to $I \propto |\chi^{ED}_{yyy}|^2 \neq 0$. For the same reason also the second peak in the χ^{MD}_{zyz} spectrum at 2.75 eV is merely a projection, corresponding to SHG from χ^{ED}_{yzz}.

Figure 6.2(b) displays further SHG spectra, this time obtained with light propagating along the z axis of MnWO$_4$. For $k\|z$, the tensor components χ^{ED}_{yyy} and χ^{ED}_{yxx} contribute to SHG, showing a pronounced spectral dependence. Explicitly strong SHG is detected from χ^{ED}_{yxx} at 1.95 eV, exceeding the χ^{ED}_{yyy} contribution by a factor of 200 (see figure 6.2(d)).

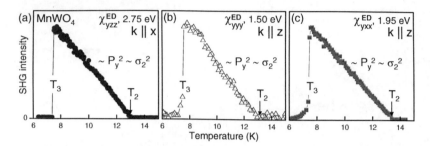

Figure 6.3: (a)-(c) Temperature-dependent measurements of the SHG intensity from χ^{ED}_{yzz}, χ^{ED}_{yyy}, and χ^{ED}_{yxx}, respectively. All contributions exhibit a linear temperature dependence, reflecting the scaling behavior of the square of the magnetic order-parameter amplitude σ_2^2 and the square of the spontaneous polarization P_y^2. The obtained SHG signals are present in the multiferroic phase only, arising at T_2 and vanishing across the first-order transition at T_3.

Apparently, the peak position does not coincide with the transition energy (2.15 eV) at which according to figure 6.1(a) a resonant excitation is expected.

However, the shift in energy and the extraordinarily strong SHG from χ^{ED}_{yxx} are merely consequences of the linear optical properties of $MnWO_4$. Both effects are hallmarks of phase matching, discussed in detail in section A.

Altogether, the spectra shown in figure 6.2 identify three independent tensor components that emerge in the multiferroic phase. The set of tensor components include χ^{ED}_{yzz}, χ^{ED}_{yyy}, and χ^{ED}_{yxx}. Their sensitivity to the phase transitions at T_2 and T_1 is evidenced by figure 6.3, presenting temperature-dependent measurements of the related SHG intensities.

Figures 6.3(a) to (c) show a linear increase of the SHG intensity below the second-order phase transition at T_2 for all three contributions. When lowering the temperature further, the SHG signals vanish again along with the multiferroicity in $MnWO_4$ across the first-order phase transition at T_3.

The obtained linear temperature dependencies in the multiferroic phase reflect the scaling behavior of the magnetic order-parameter amplitude σ_2 (η_1 is frozen in the AF2 phase) *and* of the spontaneous polarization P_y

$$I \propto |\chi^{ED}_{ijk}|^2 \propto |\sigma_2|^2 \propto |P_y|^2 \, . \qquad (6.1)$$

This follows from the previous discussion of order parameters in $MnWO_4$ and the symmetry analysis in sections 5.1.3 and 5.2. It was already pointed out that the coexisting magnetic order parameters and the ferroelectric order reduce the symmetry of the system in exactly the same way and, therefore, allow for identical SHG contributions. Moreover, equation (5.2) predicts the same scaling behavior for the primary magnetic and the secondary electric order parameter.

Thus, one can conclude that magnetic and electric SHG contributions for $k||x$ and $k||z$ are inseparably entangled. SHG from the components χ^{ED}_{yzz}, χ^{ED}_{yyy}, and χ^{ED}_{yxx} couples to the magnetic as well as to the electric subsystem and can be referred to as "multiferroic" SHG. The entanglement of magnetic and electric SHG contributions is quite remarkable and reflects the unique correlation between magnetism and ferroelectricity in magnetically induced ferroelectrics, i.e. joint-order-parameter multiferroics. In contrast, specific SHG con-

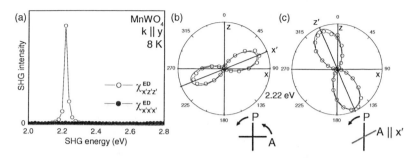

Figure 6.4: SHG contribution of magnetic origin, probed with light incident parallel to the y axis of MnWO$_4$. (a) Polarization-dependent spectra show a pronounced peak at 2.22 eV corresponding to SHG from $\chi^{ED}_{x'z'z'}$. (b, c) The rotational anisotropy of the observed SHG contribution reveals its relation to a reference system whose x' and z' axis are rotated by $\approx 25°$ with respect to the crystallographic axes x and z.

tributions coupling either to the magnetic or the ferroelectric subsystem can be separated in split-order-parameter multiferroics as hexagonal HoMnO$_3$ [129].

6.1.3 Magnetic SHG contributions

In the previous section 6.1.2, different multiferroic SHG contributions have been identified by SHG spectroscopy on the basis of the symmetry analysis in section 5.2. The analysis in particular revealed that neither ED nor MD SHG contributions are symmetry-allowed for light fields incident along the y axis of MnWO$_4$ (see table 5.2).

However, as shown in figure 6.4(a), polarization-dependent SHG spectra obtained in the multiferroic phase with $k||y$ exhibit SHG contributions around 2.22 eV. Contradictory to table 5.2, figure 6.4(a) clearly evidences the existence of a nonzero tensor component χ^{ED}_{ijk} that contributes to SHG for $k||y$.

To understand this surprising result, a polarization analysis of the SHG signal was performed. The main results are depicted in figures 6.4(b) and (c), showing the anisotropy of the SHG intensity. Figure 6.4(b) is measured by simultaneously rotating "crossed" polarizer and analyzer, whereas figure 6.4(c) is gained by rotating the polarizer only (see sketches in figures 6.4(b) and (c)).

One concludes from the anisotropy measurements that the emitted SHG light is polarized along a certain axis denoted by x' in figure 6.4(b), being induced by z'-polarized incident light fields (figure 6.4(c)). Here, the prime indicates that the polarization directions do not coincide with the crystallographic x and z axes whose orientation is also sketched in figures 6.4(b) and (c).

The two axes x' and z' can be considered as a reference system that is rotated by $\approx 25°$ with respect to the crystallographic reference system, denoted by x and z. However, the rotated reference system is not observed in SHG measurements with $k||x$ or $k||z^2$.

This indicates that the third axis y' has to coincide with the crystallographic y axis. Hence, the crystallographic principle axes and the reference system denoted by x', y',

---

[2]Anisotropy measurements obtained with $k||x$ and $k||z$ can be found in reference [148]

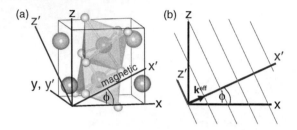

Figure 6.5: (a) Relation between the reference system defined by x', y', and z', and the crystallographic reference system (x, y, and z) in MnWO$_4$. (b) The "effective" magnetic wave vector $\mathbf{k}^{\mathrm{eff}}$ defines a set of equivalent magnetically ordered planes. These planes determine the magnetic reference system which is rotated with respect to the crystallographic reference system.

and z', appear to be related by a rotation of $\approx 25°$ around $y \equiv y'$. Their relation is schematically depicted in figure 6.5(a). The experimental results suggest the existence of a *second* reference system, that is partially decoupled from the crystallographic structure and allows for SHG from a tensor component $\chi^{\mathrm{ED}}_{x'z'z'}$. A discussion about its origin is provided in the following section.

Magnetic reference system

It is reasonable to assume that the incommensurability of the magnetic order in the multiferroic phase plays an important role. The incommensurate character may lead to additional symmetry-allowed tensor components of the nonlinear susceptibility, which is not taken into account by considering the point symmetry of the ordered phase only. This would imply that the reference system denoted by x', y' and z' can be interpreted as a "magnetic reference system".

To verify this assumption, we consider the wave vector $\mathbf{k}^{\mathrm{inc}} = (-0.241, \frac{1}{2}, 0.457)$ describing the incommensurate magnetic order. Its irrational components $k_x \approx -\frac{3}{14}$ and $k_z \approx \frac{16}{35}$ reflect a violation of translation symmtries along the x and the z axis of the crystal on a "large scale" in the magnetically ordered phases. On a "large scale" here means that a magnetic unit cell of the size $14a \geq 7$ nm, $2b \approx 1$ nm, and $35c \geq 17$ nm has to be assumed in order to obtain the translation symmetry given by $\mathbf{k}$, with lattice constants $a = 4.83$ Å, $b = 5.7603$ Å, and $c = 4.994$ Å [137,141].

The antiferromagnetic order along y cannot be resolved by optical SHG experiments. Thus, an *effective* magnetic propagation vector $\mathbf{k}^{\mathrm{eff}} = (-0.241, 0, 0.457)$ can be defined in the present case. This wave vector $\mathbf{k}^{\mathrm{eff}}$ describes a periodic arrangement of planes at intervals of ≈ 20 nm, schematically depicted in figure 6.5(b). All magnetic ions lying within one of these planes exhibit similar magnetic properties regarding magnitude $|\mathbf{M}|$ and orientation of the magnetic moments. Note that magnetic moments $+\mathbf{M}$ and $-\mathbf{M}$ are considered to be equal, since the antiferromagnetic order along y is not resolved.

The set of planes described by $\mathbf{k}^{\mathrm{eff}}$ defines two preferential directions in the xz plane: A first one which is given by $\mathbf{k}^{\mathrm{eff}}$ itself and a second one that is perpendicular to the effective wave vector. Here, the angle between $\mathbf{k}^{\text{eff}}$ and the crystallographic x axis is given by

$$\phi = \arctan\left(\frac{k_x}{k_z}\right) \approx 27.8° . \tag{6.2}$$

Apparently, the direction of $\mathbf{k}^{\text{eff}}$ is in fairly good aggreement with the direction of x' derived from figure 6.4(b).

This leads to the conclusion that the observation of the rotated axes denoted by x' and z' in figures 6.4(b) and (c) is determined by the incommensurate magnetic order in the multiferroic phase. The incommensurability leads to a violation of translation symmetries on a rather large scale that, although being in the subwavelength regime, seems to be no longer negligible in optical SHG experiments. As a consequence, a "magnetic reference system" manifests in measurements with $k\|y$. In contrast, the projections of x' and z' onto the xy plane or yz plane coincide with the direction of a crystal axis and, therefore, cannot be distinguished from them with $k\|z$ or $k\|x$, respectively (see figure 6.5(a)).

Purely optical effects as pleochroism[3] and birefringence can be excluded to be the origin for the observation of a rotated reference system. Even though birefringence may contribute to a certain extent to the rotation of the polarization direction in figure 6.4(b) and (c), effects due to birefringence should be rather weak, because a rotation of the polarization direction is not observed for $k\|x$ and $k\|z$.

In conclusion, the experimental results demonstrate that a symmetry analysis in terms of point groups only is insufficient to determine all symmetry-allowed tensor components in the presence of incommensurate order. Apparently, additional components contribute to SHG when translation symmetries are violated on a large scale by incommensurate modulations. Although no theoretical predictions for a threshold value regarding the sensitivity of the SHG process towards a violation of translation symmetries exist, the experimental results indicate that a threshold value in the subwavelength regime has to be expected.

Most remarkably, the SHG process is sensitive to the reference system defined by the incommensurate order in the sense of figure 6.5(b). Although it is known that an incommensurate quasilattice can be described by a periodic structure in a different reference system [162], no experimental observation of a partially decoupled magnetic reference system by optical SHG measurements is reported so far.

Temperature-dependent measurements

On the basis of the previous discussion SHG from $\chi^{\text{ED}}_{x'z'z'}$ can be attributed to the ordering of the magnetic subsystem. In particular, no SHG of ferroelectric origin is symmetry-allowed for $k\|y$. This suggests that the application of light fields with $k\|y$ allows for probing the order parameters of the magnetic subsystem only. In contrast, inseparably entangled magnetic and electric contributions are detected for $k\|x$ or $k\|z$.

Temperature-dependent measurements of SHG from $\chi^{\text{ED}}_{x'z'z'}$ are presented in figure 6.6. Figure 6.6(a) reveals a linear increase of the SHG intensity in the multiferroic phase below T_2, followed by an abrupt decrease at T_3, indicating the first-order AF2 $\rightarrow$ AF3 transitition. First of all, figure 6.6(a) proves that SHG from $\chi^{\text{ED}}_{x'z'z'}$ couples to the magnetic order parameter η_2, being sensitive to the magnetic phase transitions at T_2 and T_3. Thus, the measurement reflects the scaling behavior of η_2 according to

$$I \propto |\chi^{\text{ED}}_{x'z'z'}|^2 \propto |\sigma_2|^2 . \tag{6.3}$$

---

[3]One generally speaks of *pleochroism* when a crystal exhibits different absorption properties along the optical axes determined by its indicatrix [161]. In the case of MnWO$_4$ the indicatrix is a triaxial ellipsoid.

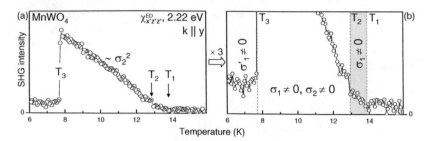

Figure 6.6: Temperature dependence of the SHG intensity from $\chi^{ED}_{x'z'z'}$ observed at 2.22 eV. (a) Between T_2 and T_3 a linear increase of the SHG intensity is observed which reflects a coupling to the active magnetic order-parameter amplitude σ_2. (b) On a closer inspection of the measurement presented in (a) nonzero SHG contributions in the AF3 and AF1 phase are identified suggesting a coupling the the order-parameter amplitudes σ_1 and σ'_1, respectively.

A comparison of figure 6.6(a) with figures 6.3(a) to (c) reveals that SHG of purely "magnetic" and SHG of "multiferroic" origin exhibit exactly the same T-dependence. This coincidence nicely confirms the predicted correlation between symmetry-breaking order parameters in MnWO$_4$ made by equation (5.2). An identical scaling behavior is observed, irrespective of probing the magnetic order parameter η_2 (figure 6.6(a)), the spontaneous polarization P_y [144], or both at the same time (figure 6.3(a) to (c)).
This is one of the unique properties of magnetically induced ferroelectrics – the spontaneous polarization behaves exactly in the same way as the magnetic order parameter that drives the multiferroic phase transition. As a consequence, the SHG field always carries information about both, the magnetic and the electric order. Thus, we can conclude that the concept of a rigorous separation in "multiferroic" and "magnetic", or even "electric" and "magnetic" SHG contributions is obsolete regarding magnetically induced ferroelectricity in MnWO$_4$. The relevant order parameters of the electric and the magnetic subsystem are inseparably entangled in the multiferroic phase, irrespective of the different microscopic origin of SHG from $\chi^{ED}_{x'z'z'}$ (magnetic), compared to SHG from χ^{ED}_{yzz}, χ^{ED}_{yyy}, and χ^{ED}_{yxx} (magnetic + electric).

6.1.4   Temperature dependence of the correlated order parameters

We now discuss the temperature dependence or "scaling behavior" of the order parameters in MnWO$_4$ in more detail. Figures 6.3 and 6.6(a) show a linear temperature dependence of the SHG intensity in the AF2 phase. The scaling behavior of the associated order parameters can be obtained from the relations (6.1) and (6.3), leading to

$$\sigma_2 \propto (T_2 - T)^{\frac{1}{2}} \quad \text{and} \tag{6.4}$$
$$P_y \propto (T_2 - T)^{\frac{1}{2}} \tag{6.5}$$

for the order-parameter amplitude σ_2 and the spontaneous polarization P_y, respectively. The results are consistent with the theoretical predictions made by Tolédano *et al.* and, therefore, confirm their assumption of a lowest order coupling term $\propto \delta\sigma_1\sigma_2 P_y$ contributing to the free energy (see section 5.1.3) [147].

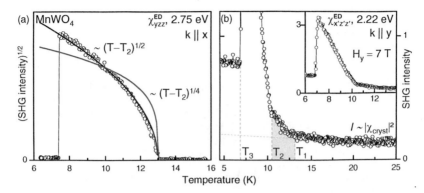

Figure 6.7: (a) To analyze the temperature dependence of the magnetic order parameter σ_2 (and P_y) the square root of the SHG intensity from χ_{yzz}^{ED} is fitted by assuming two different critical exponents. Apparently, only by assuming a critical exponent $\beta = \frac{1}{2}$ [147] the measured temperature dependence can be reproduced whereas the fit with $\beta = \frac{1}{4}$ [163] does not match the data. (b) Temperature-dependent measurement of SHG from $\chi_{x'z'z'}^{ED}$ in a magnetic field applied parallel to the y axis of the crystal. For $H_y = 7$ T the SHG contributions that occur in the magnetically ordered phases AF3 and AF1 due to magnetostriction become clearly visible (see also figure 6.6). The inset of (b) indicates that the general scaling behavior of σ_2 in the multiferroic phase is unaffected by the magnetic field.

In contrast, Lawes *et al.* obtained a different critical exponent regarding the magnetic order parameter that drives the multiferroic phase transition in Ni$_3$V$_2$O$_8$ [163]. They observed a scaling behavior according to $\sigma_{LT1} \propto (T_{LT1} - T)^{\frac{1}{4}}$ with σ_{LT1} being the relevant magnetic order-parameter amplitude that corresponds to σ_2 in MnWO$_4$. For a comparison of the results, figure 6.7(a) shows the square root of the SHG intensity

$$\sqrt{I} \propto |\chi_{x'z'z'}^{ED}| \propto |\sigma_2| \qquad (6.6)$$

measured in MnWO$_4$. In addition, two curves are presented which are gained by fitting a function $\propto (T_2 - T)^{\beta}$ to the data points using critical exponents $\beta = \frac{1}{2}$ and $\beta = \frac{1}{4}$. Apparently, only by assuming a critical exponent $\frac{1}{2}$ the measured temperature dependence is reproduced. Thus, figure 6.7(a) clearly evidences a different scaling behavior of the magnetic order parameter that drives the multiferroic transition in MnWO$_4$ compared to Ni$_3$V$_2$O$_8$, although the sequence of phase transitions in the two systems is quite similar. Apparently, the distinct microscopy of the two compounds leads to a qualitatively different scaling behavior of the inherent magnetic order parameters. Nevertheless, equation (3.14) is valid in the multiferroic phase of MnWO$_4$ and Ni$_3$V$_2$O$_8$.

Thus, on the one hand side symmetry-based theories correctly describe the general relation between the ordered spins and the induced spontaneous polarization. On the other hand, the different scaling behavior obtained points towards subtle variations regarding the microscopic properties of the two compounds which are not understood so far. The comparison of the two systems points out once more that precise microscopic models are highly-desirable for a better understanding of the complex phenomena observed in magnetically induced ferroelectrics.

Lattice-induced effects

We now come back to the temperature-dependent measurement shown in figure 6.6(a). On a closer inspection, one can see that the SHG intensity already increases below the first magnetic phase transition (T_1). In addition, a nonzero SHG contribution is observed in the AF1 phase $(T < T_3)$. This is quite remarkable, because both contributions are symmetry-forbidden for point symmetry $2_y/m_y1'$ (see table 5.2). The two contributions become even more obvious by changing the scale as shown in figure 6.6(b).

Although no SHG is symmetry-allowed for $k||y$ in the phases AF3 and AF1, the temperature-dependent measurement in figure 6.6 suggests a coupling of the SHG process to the associated magnetic order parameters η_1 and η_1', respectively.

Unfortunately, the weakness of the SHG signals hampers the analysis. Especially the contribution in the AF3 phase nearly vanishes in the noise. To solve this problem, a magnetic field of 7 T was applied parallel to the y axis of $MnWO_4$. As the (H,T)-phase diagram in figure 5.3 indicates, magnetic fields $H||y$ increase the stability range of the AF3 phase and, therefore, should significantly simplify the analysis.

Temperature-dependent measurements of the SHG intensity with $k||y$ in a magnetic field of 7 T are displayed in figure 6.7(b). In contrast to the zero-field measurement in figure 6.6(a), figure 6.7(b) now clearly reveals an increase of the SHG intensity at T_1, as well as a nonzero contribution in the magnetic ground state (AF1) while the general scaling behavior of η_2 and P_y seems to be unaffected by the magnetic field (see inset of figure 6.7(b)).

Thus, just as figure 6.6(b), figure 6.7(b) suggests that the SHG process picks up the symmetry changes imposed by the magnetic order parameters η_1 and η_1'.

However, figure 6.6(b) also clearly displays a crystallographic SHG contribution $I \propto |\chi_{cryst}|^2$ in the paramagnetic phase $(T > T_1)$. Retrospectively, this contribution can also be found in the zero-field measurement in figure 6.6(b). The SHG generated from the crystal lattice arises due to a small misalignement of the sample and naturally explains the observation of SHG in the phases AF3 and AF1:

Like all SHG signals, the crystallographic contribution originates from optically driven transitions between electronic states. Furthermore, the magnetic phase transitions in magnetically induced ferroelectrics strongly couple to the lattice parameters [164, 165]. The change in the lattice parameters is accompanied by a change in the electronic band structure. Therefore, the SHG yield from χ_{cryst} will change at the different magnetic transitions, leading to the observed temperature dependence in the phases AF3 and AF1. Thus, figures 6.6(b) and figure 6.7(b) rather reflect the impact of the nonzero magnetic order parameters on the lattice parameters than indicate a direct coupling to η_1 and η_1'. This phenomenon is generally known as magnetostriction.

The attribution of SHG in the AF3 and AF1 phase to an enhanced response from χ_{cryst} is supported by thermal expansion measurements of $MnWO_4$ [166]. The reported relative length change along the y axis of the crystal nicely correlates to the obtained temperature dependency of SHG in between T_1 and T_2, as well as below T_3.

In conclusion, nonzero SHG contributions in the AF3 and AF1 phase are of crystallographic origin, while the obtained temperature dependence is caused by the strong coupling between the magnetic ordering and the lattice parameters. This in particular means, that the emitted frequency-doubled light fields carry no information about the phase of the magnetic order parameters η_1 and η_1' which will be of importance regarding the latter investigation of the domain topology in $MnWO_4$.

6.1.5 Summary of SHG spectroscopy on MnWO$_4$

Order-parameter-related SHG contributions in MnWO$_4$ have been identified by polarization-dependent SHG spectroscopy. The presented measurements evidence that the violation of the spatial inversion symmetry by the coexisting magnetic order parameters and the induced spontaneous polarization leads to a pronounced nonlinear response of the system in the multiferroic AF2 phase. All contributing tensor components of the nonlinear susceptibility have been shown to couple to the magnetic order parameter η_2 and, via equation (5.2), to the spontaneous polarization P_y. Since even the same scaling behavior is observed, magnetic and electric SHG contributions are inseparably entangled in MnWO$_4$. This "hybrid" nature refelcts the unique correlation between magnetism and ferroelectricity in this compound.

Moreover, the observed temperature dependence of the magnetic order parameters in MnWO$_4$ was analyzed and compared to Ni$_3$V$_2$O$_8$. Although from the group-theoretical point of view the same symmetry-breaking mechanism leads to magnetically induced ferroelectricity in the two compounds, different critical exponents are obtained regarding the magnetic order parameter that drives the multiferroic transition.

6.2 Nonlinear spectroscopy of TbMn$_2$O$_5$

This section deals with the investigation of the multiferroic properties of TbMn$_2$O$_5$ by polarization-dependent SHG spectroscopy. The complex sequence of magnetic and correlated electric phase transitions in TbMn$_2$O$_5$ has already been discussed in section 5.3. In the following, various SHG contributions that couple to the symmetry-breaking order parameters are identified and analyzed with respect to their scaling behavior. Of special interest is the ferroelectric polarization in this compound which is assumed to be composed of two different components [9]. The experimental confirmation of the composite nature is still due because the pyrocurrent measurements applied so far reveal the net polarization only.

6.2.1 Multiferroic SHG contributions

In contrast to the previously discussed system MnWO$_4$, no SHG is observed in the non-multiferroic phases of TbMn$_2$O$_5$, i.e. in the paramagnetic/-electric phase and the paraelectric but magnetically ordered phase I. This drastically simplifies the analysis of multiferroicity in TbMn$_2$O$_5$ by SHG spectroscopy. One can directly conclude that all SHG contributions that are detected in the multiferroic phases II to IV couple to the active magnetic order parameter components and, via equations (5.4), (5.6), or (5.7), to the spontaneous polarization P_y. The correlation between magnetism and ferroelectricity described by equations (5.4), (5.6), and (5.7) also implies that, similar to MnWO$_4$, magnetic and electric SHG contributions are inseparably entangled and can be considered as multiferroic SHG.

According to table 5.4, SHG of multiferroic origin can be expected for incident light fields propagating either along the x or the z axis of TbMn$_2$O$_5$. To identify order-parameter-related SHG, SHG spectra were obtained in the different multiferroic phases.

Figures 6.8(a) and (b) display SHG spectra of the crystal in the multiferroic phase III (25 K) gained with $k||x$ and $k||z$, respectively[4]. All SHG contributions in figures 6.8(a)

---

[4]The relative SHG intensity is calculated by $I_{SHG}^{rel} = I_{SHG}/I_{fund}^2$ with I_{fund} being the intensity of the

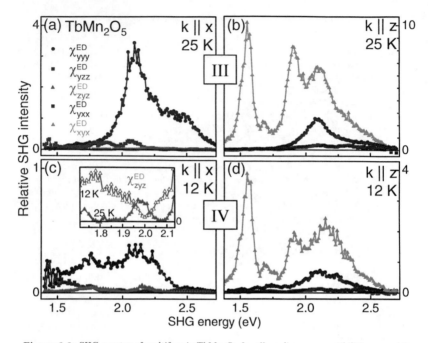

Figure 6.8: SHG spectra of multiferroic $TbMn_2O_5$ for all nonlinear susceptibilities $\chi_{ijk} \neq 0$ in the multiferroic phases III (25 K) and IV (12 K), respectively. The spectra taken with incident light fields $k||x$ and $k||z$ are shown while no SHG is observed with $k||y$. The same scale is used for SHG intensities in (a) to (d).

and (b) exhibit a pronounced spectral dependence while maxima in the SHG intensity appear at 1.55 eV, 1.9 eV and 2.1 eV. A comparison to the linear absorption spectrum of $TbMn_2O_5$ in reference [114] reveals that the related microscopic transitions correspond to the lowest O–Mn charge-transfer excitations.

The obtained tensor components χ_{ijk}^{ED} are consistent with the symmetry reduction imposed by the magnetic order parameter η_1 and the induced spontaneous polarization P_y in phase III, leading to point symmetry $m2m$ (see section 5.3.3). In particular, no SHG is detected for $k||y$.

On lowering the temperature, the spectral shape of SHG from the different components χ_{ijk}^{ED} changes. In figure 6.8(c) and (d) SHG spectra measured in the multiferroic phase IV (12 K) are presented. A comparison of the data in figures 6.8 taken at 25 K and at 12 K reveals changes in the amplitude and, even more important, a spectral redistribution of the SHG signal when going through the phase transition III $\rightarrow$ IV at T_4. The most significant change is a drop of the SHG intensity obtained from χ_{yyy}^{ED} at 2.1 eV. However, on a closer inspection, spectral regions are identified in which the SHG intensity for χ_{zyz}^{ED} is zero at

---

fundamental light field detected behind the sample. By plotting I_{SHG}^{rel} instead of I_{SHG}, the spectra are corrected for absorption effects and optical anisotropy.

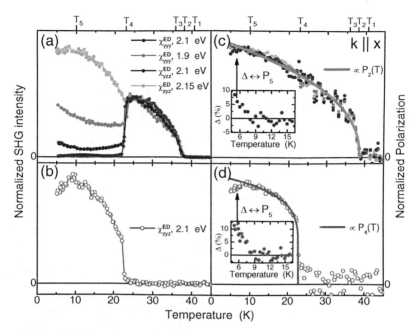

Figure 6.9: Temperature-dependent SHG intensity and spontaneous polarization in TbMn$_2$O$_5$. (a) SHG intensity for various photon energies and polarizations. (b) SHG intensity for χ^{ED}_{zyz} at 2.1 eV. (c, d) Contributions to the spontaneous polarization P_2 and P_4 extracted from (a) and (b). Straight lines are derived from fits of $I \propto (1 - T/T_{2,4})^{\beta}$. Insets show the deviation Δ between fits and data indicating the presence of a spontaneous polarization P_5.

$T > T_4$ (1.77–1.91 eV and 2.06–2.09 eV) while below T_4 a SHG signal is observed (see inset of figure 6.8(c)).

With respect to the induced polarization, this means that there are at least two independent contributions to P_y: One at $T < T_4$ that is detected in the aforementioned spectral range and one at $T > T_4$ that is not observed. Remember that all nonzero SHG contributions couple to the active components of the magnetic order parameter η_1 and, via equations (5.4), (5.6), or (5.7), to P_y.

Because of the selectivity indicated by the inset of figure 6.8(c) a separation of different contributions to P_y should be possible by SHG spectroscopy. This is the topic of the following sections.

6.2.2 Composite nature of the spontaneous polarization

The sensitivity of the SHG process to different contributions to the spontaneous polarization is exploited in figure 6.9(a) and 6.9(b) which shows the temperature dependence of the normalized SHG intensity $I(T)$ at different polarizations and photon energies. The temperature dependence at $T > T_4$ is the same for all measurements in figure 6.9(a) which indicates that a single contribution

$$I(T) \propto |a_2(2\omega)P_2(T)|^2 \quad (T > T_4)$$

to the spontaneous polarization leads to this universal scaling with a_2 as coupling coefficient. Here, the subscript $i = 2$ of a and P indicates the transition temperature T_i at which the polarization contribution arises.

Below T_4 the temperature dependence to the SHG signal is not universal anymore. At least two contributions to the spontaneous polarization are present which leads to SHG interference according to

$$I(T) \propto |a_2(2\omega)P_2(T) + a_4(2\omega)P_4(T)|^2 \quad (T < T_4) .$$

As revealed by figure 6.8 the spectral dependence of a_2 and a_4 is different which explains the non-universal temperature dependence below T_4 in figure 6.9. This is most strikingly expressed by figure 6.9(b) for which $a_2 = 0$ and $a_4 \neq 0$.

Figure 6.9(b) can be employed for separating P_2 from P_4 (and any additional contributions) in figure 6.9(a), by fitting the complex coefficient a_4 such that P_2 and $\partial P_2/\partial T$ continue steadily through T_4 [9, 92]. For this purpose, i.e. showing the universality of P_2 and its continuity at T_4, it is sufficient to fit a simplified phenomenological model assuming $I \propto (T_{2,4} - T)^\beta$. The fit function does not take into account the phase transition at T_3 at which the coupling constant δ between the magnetic and the electric order changes but fairly reproduces the experimentally observed scaling behavior. This leads to the temperature dependence of P_2 and P_4 shown in figures 6.9(c) and 6.9(d), respectively. The universality of these data sets shows that no other contributions than P_2 and P_4 are present in $TbMn_2O_5$ above T_5.

At T_5 the data begin to deviate from the power law fitted to them. This indicates the presence of a third, yet unclaimed contribution to the net polarization emerging at T_5 (The insets in figures 6.9(c) and 6.9(d) reveal $T_5 = (9.3 \pm 0.9)$ K.). The measurements suggest that the third contribution to the net polarization P_y is related to the long-range order of the Tb spins which occurs at T_5.

However, as mentioned above the SHG process is related to O–Mn charge-transfer transitions. Thus, it is reasonable to assume that the ordering of the 4f moments indirectly affects the charge-transfer transitions via magnetic exchange interactions of the Mn^{3+} or Mn^{4+} moments and those of Tb^{3+}.

In conclusion, SHG allows to uniquely identify and separate three independent contributions to the spontaneous polarization of multiferroic $TbMn_2O_5$. Moreover, three instead of the so far claimed two contributions to the net polarization are observed. The third component is attributed to the rare-earth magnetism.

An active role of the rare-earth magnetism on magnetically induced ferroelectricity was recently observed in orthorhombic $DyMnO_3$ [167]. Simlar to $TbMn_2O_5$, the ordering of the rare-earth spins enhances the spontaneous polarization in $DyMnO_3$. Figure 6.9 provides experimental evidence that a pronounced interplay of rare-earth magnetism and ferroelectricity is also present in the RMn_2O_5 compounds.

6.2.3   Scaling behavior of the ferroelectric order parameter

We now come to a detailed discussion of the scaling behavior of the magnetically induced ferroelectricity in $TbMn_2O_5$. Figure 6.10 presents an analysis of the different ferroelectric

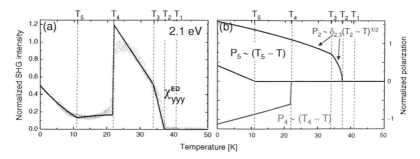

Figure 6.10: Analysis of the temperature dependence of the ferroelectric order parameter of TbMn$_2$O$_5$ in the various multiferroic phases II to IV. (a) Temperature-dependent SHG intensity from χ_{yyy}^{ED}. The straight line is derived from a fit of the data according to equation (6.7). (b) Decomposition of the three different polarization components P_i that lead to the obtained temperature dependence of the net polarization in TbMn$_2$O$_5$.

contributions, taking into account the theoretically predicted scaling behavior of P_y for the multiferroic phases II to IV [92]. Figure 6.10(a) shows a temperature-dependent measurement of SHG from χ_{yyy}^{ED} at a photon energy of 2.1 eV. One can see a linear increase of the SHG intensity at T_2 to T_3 and T_3 to T_4 while, consistent with the theory, the slope changes at T_3 due a change in the coupling constant between the magnetic order and P_y. Furthermore, an abrupt decrease of the SHG intensity at T_4 indicates the first-order III → IV transition. The intensity increases again below the ordering temperature (T_5) of the Tb moments.

On the basis of the previous discussion, the data in figure 6.10(a) can be fitted by assuming three independent contributions to the net polarization according to

$$I \propto |P_2 + P_4 e^{i\phi} + P_5 e^{i\phi'}|^2 \,, \tag{6.7}$$

with P_i denoting the different contributions to P_y that arise at T_i. ϕ and ϕ' parametrize the relative phases of the associated SHG signals.

The red curve in figure 6.10(a) shows the resulting fit which was gained by taking into account the specific scaling behavior of the three contributions to P_y (see section 5.3.3)

$$P_2 \propto \delta_{2,3}(T_2 - T)^{\frac{1}{2}} \tag{6.8}$$
$$P_4 \propto (T_4 - T) \tag{6.9}$$
$$P_5 \propto (T_5 - T) \,, \tag{6.10}$$

with $\delta_{2,3}$ reflecting different coupling constants for phase II and III. Apparently, the fitted function (6.7) nicely reproduces the experimental observation, evidencing the presence of three independent polarization contributions. Only around T_4 a deviation from the theoretical model is observed which is merely a consequence of a non-uniform temperature distribution over the sample, leading to a phase coexistence of phase III and phase IV.

Most important, by fitting the SHG data the three contributing polarization components P_i to P_y can be decomposed. The theoretically derived individual contributions are displayed in figure 6.10(b). One can see that the first magnetically induced contribution P_2 arises at T_2 showing a pseudo-proper scaling behavior, i.e. scales like the primary magnetic order parameter. The same scaling behavior is obtained at $T < T_3$ – only the coupling constant changes at the transition ($\delta_2 \rightarrow \delta_3$).

At T_4 the second contribution P_4 emerges through a first-order phase transition. The polarisation P_4 points in the opposite direction of P_2, leading to the decrease of the SHG intensity in figure 6.10(a) at T_4. In contrast to P_2, P_4 exhibits a linear scaling behavior which is characteristic for improper ferroelectricity [46].

Finally, the third contribution P_5 shows up at T_5 as the Tb 4f spins order. This rare-earth induced contribution P_5 is of the same orientation as P_2 and leads to the enhancement of the spontaneous polarization P_y at $T < T_5$. The best fit of the data in figure 6.10(a) is achieved by assuming $P_5 \propto (T_5 - T)$ which expresses the improper nature of P_5.

In conclusion, figure 6.10 clearly demonstrates the presence of three independent contributions to the net polarization in TbMn$_2$O$_5$. Note that the concept of three contributions to P_y is qualitatively different from the model proposed by Hur *et al.* in reference [9] where only two ferroelectric "components" are considered. In the latter case the observed increase of P_y below $T = 10$ K is attributed to a decrease of P_4. In contrast, figure 6.10 suggests universal scaling of P_4 and a rare-earth induced component P_5 that leads to the enhancement of P_y. In particular, the model proposed by Hur *et al.* fails to explain the experimental results obtained by optical SHG.

6.2.4   Summary of SHG spectroscopy on TbMn$_2$O$_5$

Polarization-dependent SHG spectroscopy was applied to identify SHG contributions that couple to the symmetry-breaking order parameters in TbMn$_2$O$_5$. It has been shown that, similar to MnWO$_4$, SHG of magnetic and electric origin is inseparably entangled in this compound, providing simultaneously access to both the magnetic and the ferroelectric order in the different multiferroic phases.

The analysis of the measured SHG spectra and temperature-dependent measurements has uncovered three independent contributions to the net spontaneous polarization in TbMn$_2$O$_5$ whose superposition leads to the obtained T-dependence in integrating pyroelectric measurements. By fitting the experimental SHG data, the three different polarization contributions have been separated and their universal scaling behavior has been revealed.

6.3   Microscopic origin of the magnetically induced polarization

Among other aspects, the scaling behavior of the magnetically induced ferroelectric polarization in MnWO$_4$ and TbMn$_2$O$_5$ was discussed in the previous sections 6.1 and 6.2. However, so far nothing was said about the microscopic origin of the spontaneous polarization. In both systems the polarization forms as a consequence of the magnetic order. Thus, either an ionic displacement, an electronic redistribution, or a combination of the two may be the origin for $P_y \neq 0$ (see section 3.3). The experimental verification of one or the other is highly desirable for the two compounds but hampered by the smallness of the spontaneous polarization.

In the following, SHG measurements are discussed that allow an inaugural experimental statement on the controversial theoretical discussion on the nature of the multiferroic polarization in magnetically induced ferroelectrics as MnWO$_4$ and TbMn$_2$O$_5$ [30, 95, 103, 107, 114–117].

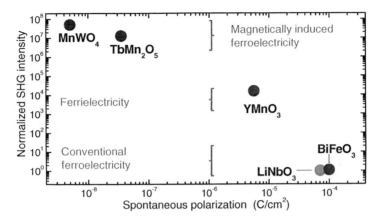

Figure 6.11: Normalized SHG yield for various joint- and split-order-parameter multiferroics and a textbook ferroelectric. Data points were derived as detailed in the text.

Figure 6.11 displays normalized SHG intensities, i.e. the SHG yield, obtained from various types of spontaneously polarized materials.

The two joint-order-parameter multiferroics $MnWO_4$ and $TbMn_2O_5$ are chosen as model compounds for systems with magnetically induced ferroelectricity. While antisymmetric spin-spin exchange interactions drive the multiferroicity in $MnWO_4$ mainly symmetric spin-spin interactions participate in driving the multiferroic state in $TbMn_2O_5$ [96, 103, 114]. $MnWO_4$ and $TbMn_2O_5$ are contrasted to three ionic ferroelectrics which are model systems for ferroelectricity driven by electrostatic effects ($YMnO_3$), by electronic lone pairs ($BiFeO_3$), and by hybridization, respectively. In addition, $YMnO_3$ [84,168] and $BiFeO_3$ [169] are split-order-parameter multiferroics while $LiNbO_3$ is a textbook ferroelectric without accompanying magnetic order.

The SHG yield of the various compounds is obtained by calculating

$$I_{SHG}^{norm} = \frac{I_{SHG}^{rel}}{P_{max}^2 \cdot d}$$

and plotting its maximum value in the range from 1.2 to 3.2 eV ($I_{SHG}^{rel} \propto |P|^2$). Here P_{max} is the maximum value of the spontaneous polarization and d is the depth of the sample up to which a contribution to the net SHG yield occurs. The calculated values of d are summarized in table 6.1. They were determined including absorption as well as phase-matching effects. Thus, I_{SHG}^{norm} in figure 6.11 represents the efficiency of the coupling between the spontaneous polarization and the SHG yield.

The comparison of different ferroelectric systems in figure 6.11 reveals an extraordinarily strong nonlinear response from magnetically induced ferroelectricity. The efficiency of the SHG process in $MnWO_4$ and $TbMn_2O_5$ appears to be three to seven orders of magnitude higher than for the other compounds irrespective of the very different microscopy of the to compounds.

Hence, regarding the SHG yield, the joint-order-parameter multiferroics clearly separate

	Signal depth d	SHG energy	Polarization
$MnWO_4$	0.46 μm	2.75 eV	5 nC/cm^2
$TbMn_2O_5$	≈ 1 μm	2.2 eV	40 nC/cm^2
$YMnO_3$	0.22 μm	2.78 eV	5.6 μC/cm^2
$LiNbO_3$	80.29 μm	2.27 eV	71 μC/cm^2
$BiFeO_3$	in reflection	2.92 eV	100 μC/cm^2

Table 6.1: Overview of the spontaneous polarization values and the depth d of the samples up to which a contribution to the net SHG yield occurs [170,171]. Due to strong absorption in the spectral range from 2 to 3 eV, the nonlinear response of $BiFeO_3$ was measured in reflection, using a femto-second laser system.

from the split-order-parameter multiferroics $YMnO_3$ and $BiFeO_3$ and also from the ferro-electric only compound $LiNbO_3$.

The huge difference in the SHG yield between the magnetially-induced ferroelectrics and the ionic ferroelectrics can be understood considering the microscopy of the SHG process. Microscopically, SHG originates from optically-driven transitions between electronic states. Because of this electronic nature SHG is particularly sensitive to electronic (in comparison to ionic) contributions to ferroelectricity. This sensitivity will be further enhanced if transitions involving the 3d band are probed whose electrons are responsible for the magnetic order and any acentric redistributions of the electron density following from it. In contrast, the coupling of nuclear displacements to the light field is small because it is based on secondary shifts of the electron cloud induced by the nuclear displacement. This means that a coupling of the light field to a primary non-centrosymmetric redistribution of the electron cloud can be orders of magnitude more efficient.

Thus, the huge difference strongly points to an electronic nature of the spontaneous polarization in $TbMn_2O_5$ and $MnwO_4$. The small value of the spontaneous polarization in these joint-order-parameter multiferroics is compensated by the giant efficiency of the SHG process so that spontaneous polarizations down to $10 - 100$ pC/cm^2 can be detected.

It is remarkable that irrespective of the aforementioned very different microscopy and composition of the spontaneous polarization in $TbMn_2O_5$ and $MnWO_4$ their normalized SHG yield is very similar. With all the aforementioned differences between the two compounds, this is a strong hint that the electronic polarization is common to joint-order-parameter multiferroics in general. Apparently, electronic contributions to the spontaneous polarization occur whenever breaking of the symmetry by magnetic order induced ferroelectricity. The placement of the split-order-parameter multiferroics in figure 6.11 is also quite noteworthy. $BiFeO_3$ is an incommensurate spin-spiral multiferroic like $MnWO_4$ and, in part, $TbMn_2O_5$. However, ferroelectricity is caused by nuclear displacement with lone-electron-pair formation and occurs independent of the magnetic order. This clearly places the compound among the ionic polar materials in figure 6.11.

$YMnO_3$ yields the largest value of I_{SHG}^{norm} among the ionic ferroelectrics. Its spontaneous polarization is 5.6 μC/cm^2 but composed of nearly compensating sublattice polarizations that are much larger [84]. Just as in figures 6.9(b) and the inset in figure 6.8(a) there will be spectral regions in which SHG is sensitive to one of the sublattices only, reproducing a much higher polarization than the net value of 5.6 μC/cm^2. This requires renormalization of the data point of $YMnO_3$ to lower values and enhances the separation between SHG from ionic and electronic contributions to the spontaneous polarization.

In conclusion, the comparison of the SHG yield of different ferroelectrics in figure 6.11

reveals a clear separation of magnetically induced ferroelectrics from ionic ferroelectrics. The extraordinarily pronounced coupling of the SHG process to the magnetically induced ferroelectricity in $MnWO_4$ and $TbMn_2O_5$ strongly points towards an electronic nature of the spontaneous polarization.

Chapter 7

Domain structure in magnetically induced ferroelectrics

This chapter deals with the complex domain topology in magnetically induced ferroelectrics. The main part of this thesis has been devoted to the analysis of the magnetic and electric domains in this challenging class of materials by SHG topography. For the first time, the inherent domain structure of a magnetically induced ferroelectric has been resolved, using MnWO$_4$ as a model system. The main experimental results are presented and discussed in this chapter, covering measurements of the full three-dimensional domain topology and its response to a controlled manipulation. By manipulating the domain structure through thermal annealing procedures or by application of external magnetic/electric fields, correlation effects are revealed and the driving forces that lead to the formation of the obtained domain topology are uncovered. Moreover, a topological magnetoelectric memory effect is reported that allows one to reconstruct the entire multiferroic domain state subsequent to quenching it. Parts of the results are already publised in Meier *et al.*, PRL **102**, 107202 (2009) or Meier *et al.*, PRB **80**, 224420 (2009).

7.1   Domains in MnWO$_4$

In the introductory sections it was pointed out that the intrinsically strong magnetoelectric interactions in magnetically induced ferroelectrics render these systems interesting for multifunctional device design [16]. The magnetoelectric coupling between the ordered spins and the electric properties in this class of materials enables magnetic phase control by application of an electric field or, vice versa, the switching of a spontaneous polarization by a magnetic field (see chapter 3).

Of particular interest in this context are the magnetic and electric domains in magnetically induced multiferroics, because, at its root, any magnetoelectric interaction originates from their correlation. Therefore, understanding the intrinsically strong magnetoelectric effects in systems with magnetically induced ferroelectricity seems synonymous to understanding the nature and interactions of their multi-domain state. However, although domains are known to be present in these systems, conventional techniques fail to image the inherent domain topology. Often an antiferromagnetic ordering of the spins emerges in these compounds so that no macroscopic magnetic moments exist that can be exploited to image the magnetic domain structure. Moreover, the small value of the spontaneous polarization (1–10 nC/cm^2) prohibits the application of established imaging techniques such

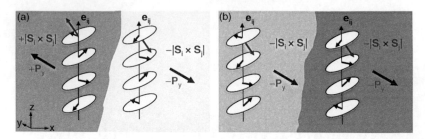

Figure 7.1: Different types of multiferroic domains in MnWO$_4$. (a) Absolute domains are
defined by the chirality $S_i \times S_j$ of the elliptical spin spiral, coupled 1 : 1 to the direction of
the spontaneous polarization P_y. (b) Translation domains (relative domains) differ in phase
regarding the spin-spiral arrangement. Domain walls between translation domains have to be
considered as a discontinuity in the periodic modulation of the spins, whereas chirality and
direction of P_y remain unaltered across the wall.

as piezoresponse force microscopy (PFM) for mapping the ferroelectric domains. Thus,
nearly nothing is known about the domain topology of these systems. So far, experiments
have been resticted to indirectly show the presence of domains in these compounds while
the prime target was actually to *remove* these domains by converting the samples to or in
between single-domain states [25, 26, 39–41].

In the following we focus on the domain topology in MnWO$_4$. Compared to TbMn$_2$O$_5$, this
compound has a relatively simple phase diagram with only one multiferroic phase in zero
magnetic field but nevertheless has great flexibility for displaying magnetoelectric coupling
effects because of the complexity on the Mn^{2+} spin spiral.

Regarding the domain structure of MnWO$_4$ in the multiferroic AF2 phase, two different
types of domains are expected. They are schematically depicted in figures 7.1(a) and
(b). Figure 7.1(a) shows domains of opposite chirality and, according to equation (3.14)
and (5.2), of opposite polarization. Because of the frozen state of η_1 only the magnetic order
parameter η_2 and the direction of P_y reverse across related domain walls. As pointed out
in sections 6.1.2 and 6.1.3, the magnetic and the electric order parameter are inseparably
entangled. Thus, the domains sketched in figure 7.1(a) can be referred to as "absolute"
domains because the mutual relation of these domains is independent from the choice of
the reference system and the value and direction of the order parameter is uniquely defined.

However, due to the periodic modulation of the spin structure described by the magnetic
order parameters η_1 and η_2, a second type of domains emerges in the multiferroic phase.
These domains are magnetic translation domains, sketched in figure 7.1(b). Different magnetic translation domains (i) and (j) exhibit the same chirality and direction of P_y whereas
the elliptical spin-spirals of different domains are out of phase by $\Delta\psi = \theta_2^{(j)} - \theta_2^{(i)}$ considering the associated complex magnetic order parameters (see also sections 3.4 and 5.1.3).
Thus, the domain walls between translation domains represent discontinuities in the modulation of the spin arrangement. These discontinuities evolve when independent nuclei
of the AF2 phase grow and converge on lowering the temperature. Different nuclei are
described by different phase angles ψ and, therefore, arbitrary values of $\Delta\psi$ appear at
domain walls between magnetic translation domains.

Alternatively, magnetic translation domains may be termed as "relative" domains because a reference system (another translation domain) is needed to distinguish the domains and associate a phase angle to the order parameter.

In summary two fundamentally different types of domains are expected to emerge in the multiferroic phase of MnWO$_4$: The absolute domains that differ by 180° in the orientation of the associated order parameter, and the relative domains, corresponding to a different phase regarding the associated magnetic order parameter. None of the two domain types can be imaged by conventional techniques probing magnetism. While at least the population of the two possible absolute domain states in spin-spiral multiferroics can be probed by neutron scattering [41], the translation domains are completely invisible.

Note that an arbitrary number of relative domains may arise in the AF2 phase of MnWO$_4$. In contrast, only two distinct absolute domains are symmetry-allowed. This follows from the fact that the relative domains result from the aforementioned independent growth of nucleation centers of the AF2 phase, whereas the absolute domains emerge, because the point symmetry is reduced at the AF3 → AF2 transition (T_2).

Whenever the point symmetry is reduced at a ferroic phase transition, the number of equivalent ferroic domain states N in the ordered phase is given by the number of symmetry operations n_{PARA} belonging to the high-temperature point group divided by the number of symmetry operations n_{AF2} of the ordered phase [48]

$$N = \frac{n_{\mathrm{PARA}}}{n_{\mathrm{AF2}}} . \tag{7.1}$$

Considering MnWO$_4$, eight point-symmetry operations belong to the high-temperature point group $2_y/m_y1'$, i.e.

$$1, \overline{1}, 2_y, \overline{2}_y, \underline{1}, \underline{\overline{1}}, \underline{2}_y, \underline{\overline{2}}_y .$$

The multiferroic phase has point symmetry $2_y1'$ ($n_{\mathrm{AF2}} = 4$) with the point-symmetry operations

$$1, 2_y, \underline{1}, \underline{2}_y ,$$

leading to $N = 2$. Thus, two different absolute domains are expected to emerge in the multiferroic AF2 phase.

Remarkably, SHG topography allows for imaging both, the absolute *and* the relative domains in spin-spiral multiferroics, because SHG is sensitive to the phase of the associated order parameters as discussed in detail in sections 4.4.2 and 4.4.3.

Experimentally resolved domains

To provide a general overview, figures 7.2(a) to (c) display spatially-resolved SHG measurements taken within the different ordered phases of MnWO$_4$ and a schematic illustration of the associated magnetic order. The images are gained on an x-oriented sample with a thickness of 60 μm using SHG from χ_{zyz}^{MD} at 2.15 eV (figures 7.2(a) and (c)) and χ_{yzz}^{ED} at 2.75 eV (figure 7.2(b)). While figures 7.2(a) and (c) reveal a homogeneous distribution of the SHG intensity in the phases AF3 and AF1, about ten curved black lines on an otherwise homogeneous background are observed in the multiferroic AF2 phase (figure 7.2(b)).

Such lines are a hallmark for the presence of absolute domains with opposite orientation of the corresponding order parameter (see figure 4.4). Because of the linear coupling to the order parameter, the SHG light field experiences a sign reversal corresponding to a 180° phase shift when crossing the domain wall. This leads to destructive interference in the vicinity of the domain wall and, therefore, to the black lines as detailed in section 4.4.2. Figure 7.2 reveals that such 180°–domains are present in the multiferroic AF2 phase only

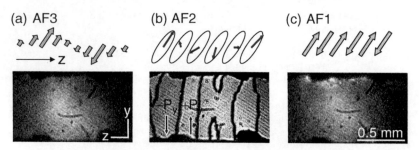

Figure 7.2: Schematic illustration of the magnetically ordered phases in MnWO$_4$ and spatially resolved SHG measurements. (a) Below the Néel temperature T_1 the Mn^{2+} moments display collinear incommensurate long-range order within the easy plane. (b) At T_2 an additional transverse spin component orders which leads to a helical incommensurate arrangement of spins. (c) In the magnetic ground state below T_3 a commensurate antiferromagnetic spin ordering is obtained. The spin spiral in the AF2 phase induces a spontaneous electric polarization along the y axis (indicated by the red arrows in (b)) while the AF1 and AF3 phases are not multiferroic. The SHG images show that a domain structure (with walls indicated by the black lines) is detected in the multiferroic AF2 phase only.

while being absent in the AF1 and AF3 phase. These 180° domains are the aforementioned absolute domains, corresponding to opposite magnetic order parameter amplitudes σ_2 and, according to equation (5.2), an opposite spontaneous polarization P_y which is indicated by the red arrows in figure 7.2(b).

This observation can be understood regarding the related point symmetry (see section 5.1.3). The point symmetry of the paramagnetic/-phase ($2_y/m_y1'$) is preserved by the commensurate up-up-down-down spin configuration of the AF1 phase and also by the incommensurate spin-density wave in the AF3 phase, because the magnetic order breaks translation symmetries only. Therefore a formation of absolute domains is symmetry-forbidden and should indeed not occur.

However, provided that χ_{zyz}^{MD} couples to the magnetic order parameter η_1 and η_1', at least a signature of the relative domains is expected in figures 7.2(a) and (c), respectively, because translation symmetries are violated by the magnetic order. Nevertheless, spatially-resolved measurements of SHG from χ_{zyz}^{MD} with $k||x$ (or $\chi_{x'z'z'}^{\mathrm{ED}}$ with $k||y$) generally exhibit a homogenous distribution of SHG intensity in the phases AF1 and AF3. This leads to the conclusion that neither SHG from χ_{zyz}^{MD} nor $\chi_{x'z'z'}^{\mathrm{ED}}$ directly couples to the magnetic order parameter η_1' or η_1, i.e. carries information about their phase.

Thus, figures 7.2(a) and (c) further confirm the results from the spectroscopy measurements in section 6.1 which already indicated that SHG from χ_{zyz}^{MD} and $\chi_{x'z'z'}^{\mathrm{ED}}$ in the AF3 and AF1 phase is generated by the crystal lattice. As a consequence, no domain structure is observed when SHG from χ_{zyz}^{MD} or $\chi_{x'z'z'}^{\mathrm{ED}}$ is probed.

Therefore we will focus on the domain structure in the multiferroic AF2 phase in the following.

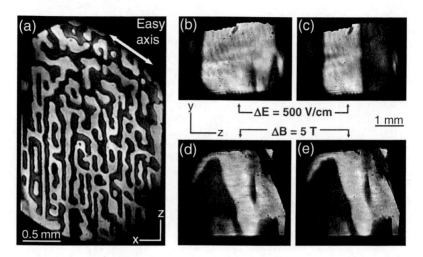

Figure 7.3: Spatially resolved SHG images reveal the hybrid nature of the multiferroic domains in MnWO$_4$: (a) Domain structure in the xz plane. A pronounced elongation along the magnetic easy axis of the crystal (yellow arrow) and the bubble topology reflect the magnetic aspect of the domains. (b, c) Domain structure in electric fields of (b) 1.67 kV/cm and (c) 2.17 kV/cm applied along the y axis. (d, e) Domain structure in magnetic fields of (d) 0 T and (e) 5 T applied along the x axis. Only the electric field affects the distribution of the domains, thus reflecting their ferroelectric aspect.

7.2 Hybrid-multiferroic (absolute) domains

First of all, the topology of the absolute domains in the multiferroic AF2 phase and their response to external poling fields will be investigated. For this purpose, the different multiferroic SHG contributions identified in section 6.1 can be employed[1].

Figure 7.3(a) shows the distribution of the absolute domains in the xz plane after zero-field cooling, imaged with SHG from $\chi_{x'z'z'}^{ED}$ on a y-oriented sample with a thickness of $d = 400\mu$m (see also figure 6.4).

We find a labyrinth-like arrangement of dark and bright areas, distributed in approximately equal proportions across the sample. The different level of brightness corresponds to opposite magnetic vector chirality and spontaneous polarization according to equation (3.14).

In contrast to figure 7.2(b), opposite domains exhibit a different level of brightness (in addition to the mere domain walls) because of the interference of the SHG light field arising from the multiferroic order with a homogeneous SHG light field generated by the crystal lattice[2] with $I \propto |\chi_{\text{cryst}}|^2$. The topology of the domains in figure 7.3(a) reveals two preferential directions: along the z axis and along a line including an angle of about 34° with the x axis.

---

[1] The set of independent tensor components contributing to order-parameter-related SHG in the AF2 phase is $\chi_{x'z'z'}^{ED}$, χ_{yzz}^{ED}, χ_{yyy}^{ED}, and χ_{yxx}^{ED} (see figures 6.3 and 6.6)

[2] Since the particular polarization dependence of crystallographic SHG contributions is irrelevant regarding the topology of the domains, the simplified notation χ_{cryst} is used in the following.

Figure 7.3(a) clearly expresses the magnetic origin of the domain structure. First, the texture of the domains strikingly resembles the patterns of bubble and stripe domains, universally attributed to modulated phases which are stabilized by competing interactions [172]. In MnWO$_4$ competing long-range magnetic interaction stabilize the modulated AF2 phase and the periodicity is determined by the magnetic wave vector $\mathbf{k}^{\text{inc}}$. Second, the preferential direction of the domain walls parallel to the arrow in figure 7.3(a) is in striking coincidence with the direction of the magnetic easy axis of MnWO$_4$ which encloses an angle of $34° - 37°$ with the x axis in the xz plane [141, 143].

In figures 7.3(b) to (e) the response of the domain structure in the yz plane to electric and magnetic fields is shown, imaged with SHG light from χ^{ED}_{yzz} on an x-oriented sample ($d = 830\mu$m). Again, opposite domains are distinguished by the different level of brightness.

It is obvious that with an electric field applied along the y axis small changes in the order of 1 kV/cm lead to pronounced changes in the topology of the absolute domains. In contrast, a magnetic field of up to 5 Tesla applied along the x axis does not alter the domain structure at all (figures 7.3(d) and 7.3(e)), and the same holds for fields applied along the y or z axis (not shown). Figures 7.3(b) to (e) thus emphasize the electric nature of the AF2 domain structure.

This leads to the conclusion that the absolute domains exhibit hallmarks of both a magnetically *and* an electrically ordered state. The *topology* is entirely determined by the *magnetic* character of the domains, whereas the *field response* reflects the *electric* character.

Because of this bilateral nature, only a denomination of the absolute domains as *hybrid-multiferroic domains* seems appropriate, whereas a description in terms of ferroelectric or antiferromagnetic domains remains incomplete. Apparently, the rigid coupling between the primary magnetic and the secondary (improper) electric order parameter pointed out by equation (5.2) also manifests in the domain topology of the multiferroic state. However, figure 7.3 provides first experimental evidence that this relation is valid on the level of domains. Only this leads to the realization that an extended concept of multiferroic hybrid domains is required.

The full three-dimensional topology of the multiferroic (absolute) domains is presented in figure 7.4 for three differently oriented MnWO$_4$ samples obtained from the same batch[3]. Figure 7.4(a) shows the domain structure in the xz plane.

Like in figure 7.3(a), dark and bright regions correspond to the two possible multiferroic domains with opposite order parameters. One can see the aforementioned elongation of domains along the magnetic easy axis, with the spontaneous polarization P_y pointing into and out of the plane. The propagation of domain walls along the direction of the spontaneous polarization is revealed by figures 7.4(b) and 7.4(c).

Interestingly, the domain walls, here again indicated by the black lines, continue rather straight along the y axis of the crystal. Apparently, the domains tend to form platelets in planes defined by the magnetic easy axis and the direction of the spontaneous polarization. The lateral extension of the domains can be quantitatively described by the "domain width" defined by

$$\text{Domain width} = \frac{2}{\pi} \cdot \frac{\text{Total test line length}}{\text{Number of intersections}} . \tag{7.2}$$

in reference [119]. This stereographical method allows to determine the domain width of an arbitrary domain pattern by counting the intersections of domain walls with arbitrary test lines as indicated for one "test line" in figure 7.3(a). In this case twelve intersections with

---

[3]While the y-oriented sample has a thickness of $d = 400\mu$m, the samples oriented along x and z have a thickness of 60μm and 90μm, respectively.

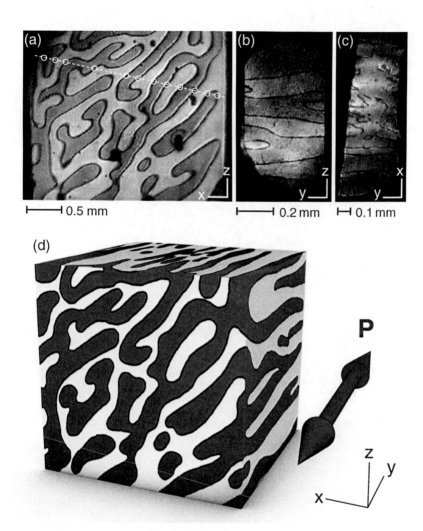

Figure 7.4: Three-dimensional distribution of the multiferroic domains. (a – c) Domains in the xz, yz, and xy plane of MnWO$_4$ samples, taken from the same batch. SHG images are gained with SHG light from $\chi^{ED}_{x'z'z'}$, χ^{ED}_{yzz}, and χ^{ED}_{yxx}, respectively. (d) Three-dimensional visualization of the multiferroic domain structure in (a) to (c).

domain walls are identified. The evaluation of twenty arbitrary test lines leads to a lateral extension of 170 ± 60 μm regarding the domains in figure 7.4(a) which is a typical value for spiral and non-spiral antiferromagnetic structures (usually 0.1 to 1 mm [124,173]).

For an improved visualization of the anisotropic multiferroic domain structure, figures 7.4(d) shows a three-dimensional simulation based on the distribution of domains in figures 7.4(a) to (c). The respective domain structure was projected onto three faces of a cuboid and mended minimally at the edges.

The three-dimensional sketch nicely visualizes the anisotropic character of the multiferroic hybrid domains and emphasizes once again the existence of preferred directions regarding the formation of domain walls.

In conclusion, figures 7.2 to 7.4 revealed the structure of the absolute domains in the multi-ferroic phase of $MnWO_4$, i.e. the domain topology in the magnetically induced ferroelectric phase. It was shown that the rigid coupling of the inseparably entangled magnetic (η_2) and electric (P_y) order parameters manifests on the level of domains.

However, it was already mentioned, that additional magnetic degrees of freedom exist, giving rise to the formation of magnetic translation (relative) domains so that further uncorrelated magnetic domain walls may arise. This is the topic of the following section 7.3.

7.3   Magnetic translation (relative) domains

Magnetic translation domains are expected to emerge and further subdivide the absolute domains in the multiferroic AF2 phase. As discussed before, the phase shifts between relative domains are arbitrary and may in particular have values $\ll 180°$. Thus, for visualizing coexisting relative domains within an absolute domain, a reference SHG signal is mandatory for enhancing the contrast between the domains as explained in section 4.4.2. For this purpose, either an external SHG reference signal or intrinsic crystallographic SHG contributions can be used.

In $MnWO_4$, SHG of crystallographic origin has been employed as a (inherent) reference signal. Following the notation introduced in section 4.4.2, the phase relation ψ' between the multiferroic SHG from one (arbitrarily) chosen relative domain (domain (1)) and crystallographic SHG (χ_{cryst}) is given by $\psi' = \theta_2^{(1)} - \psi_{ref}$. Here, $\theta_2^{(1)}$ denotes the phase of the associated magnetic order parameter η_2 while ψ_{ref} is the phase of the order-parameter-independent SHG light field. In the present case, ψ' is defined by the properties of the material, but it can nevertheless be tuned by changing the wave length of the incident light field or by rotating the sample in a way that additional crystallographic SHG components are admixed. The contrast between the relative domains can be controlled by adjusting the amplitudes of the two interfering light fields.

Figure 7.5 shows two SHG images of the domain structure in $MnWO_4$, gained at 2.15 eV and 2.75 eV after optimizing the contrast with respect to the domain topology obtained at 2.15 eV. For this purpose an x oriented sample with a thickness of 830 μm was rotated by $\phi = 10°$ around its z axis as depicted in figure 7.5(a). Similar to figures 7.3(b) to (e), figure 7.5(b) displays spatially-resolved measurements of SHG from $\chi_{yzz}^{ED} + \chi_{cryst}$ at a photon energy of 2.75 eV. Again, opposite absolute domains manifest as regions with a different level of brightness while minor changes in the contrast between the absolute domains are observed, due to the applied rotation.

Figure 7.5(c) presents a SHG image gained at 2.15 eV. Here, the polarization direction of the incident light field and the probed SHG were chosen by setting ϕ appropriately

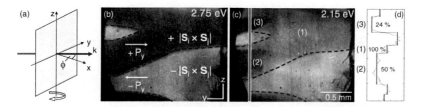

Figure 7.5: Absolute and relative domains in $MnWO_4$. (a) For imaging the multiferroic domains, an x-oriented sample is rotated around the z axis to admix an inherent SHG reference wave and thereby adjust the contrasts between different domains. (b) Absolute domains with opposite chirality $S_i \times S_j$ and polarization P_y. The domain walls are emphasized by the red dashed lines. (c) Absolute and relative domains. Three regions of different brightness are observed within one absolute domain, being related to spin spirals with a different phase $\theta_2^{(i)}$ ($i = 1, 2, 3$) of η_2. (d) Distribution of SHG intensity along the green line in figure 7.5(c). A jump in the intensity indicates the position of a domain wall.

such that contributions from χ_{cryst} (mainly χ_{zyz}^{MD}) and χ_{yyy}^{ED} are approximately of the same amplitude. As indicated by equation (4.20), equal amplitudes lead to maximal contrast between domains with a different phase of the magnetic order parameter η_2.

After optimizing the contrast, SHG topography clearly reveals a substructure in one of the absolute domains, indicated by the different level of brightness. Three regions are identified and denoted by (1), (2), and (3) in figure 7.5(c). A comparison with figure 7.5(b) evidences that the domains (1) to (3) are of the same chirality and orientation of P_y. However, the different levels of brightness reflect that the three domains differ regarding the phase of the associated order parameter η_2 (see figure 7.5(d)). These domains are the relative domains. Hence, figure 7.5 evidences the presence of two different types of domains in the multiferroic phase of $MnWO_4$, namely absolute and relative domains, alias hybrid-multiferroic and magnetic translation domains.

As mentioned before, only two types of absolute domains exist, whereas an arbitrary number of relative domains may emerge in the multiferroic phase. This symmetry-based prediction is supported by figure 7.6, showing absolute and relative domains obtained in consequtive cooling cycles. One can see in figures 7.5(b) and 7.6(a ,c) that, consistent with the symmetry analysis, only two different types of absolute domains are observed. In contrast, the number of relative domains is arbitrary and changes for each applied cooling cycle as indicated by figures 7.5(c) and 7.6(b ,d).

Thus, we can conclude that relative domains are present in the multiferroic phase of $MnWO_4$ and further subdivide the structure of the hybrid-multiferroic (absolute) domains. Since the domain walls between relative domains do not couple to the walls between the absolute domains, they constitute an additional magnetic degree of freedom. To what extend this magnetic degree of freedom can be controlled will be investigated in section 7.4.2.

7.4   Manipulation of multiferroic domains

In the following the "dynamic" aspects in the formation of the different multiferroics domains are considered. First of all, the effect of temperature annealing in an order ↔ disorder cycle is discussed on the basis of domain images gained before and after cyclig the

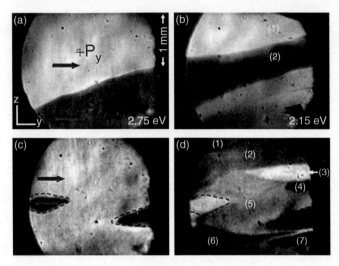

Figure 7.6: Absolute and relative domains within the yz plane of MnWO$_4$. Images were gained at 8 K with SHG at (a, c) 2.75 eV from $\chi^{\mathrm{ED}}_{yzz}+\chi_{\mathrm{cryst}}$ and (b, d) 2.15 eV from $\chi^{\mathrm{ED}}_{yyy}+\chi_{\mathrm{cryst}}$ in consecutive cooling cycles. To optimize the contrast between the relative domains observed at a photon energy at 2.15 eV, the sample was rotated around the z axis by $\phi = 10°$ as depicted in figure 7.5(a). While two types of absolute domains, i.e. domains of opposite chirality and polarization P_y, arise in the multiferroic phase (a, c), an arbitrary number of relative domains emerges (b, c).

temperature. In the second step, the possiblity of an electric-field control of the mulitferroic state is investigated, whereas topological magnetoelectric memory effects are discussed in the last part of this section.

7.4.1 Thermal annealing

A comparison of the local domain structure in subsequent measurements can provide valuable information regarding the driving forces for the formation of a domain state. Here, the effect of temperature-annealing in an order $\leftrightarrow$ disorder cycle from the AF2 phase to the paramagnetic state above T_1 and back to the multiferroic phase is discussed which is schematically illustrated in 7.7(a). Initially, the sample was zero-field-cooled from room temperature into the AF2 phase which leads to the SHG image shown in figure 7.7(b).

Figure 7.7(b) displays the multiferroic hybrid (absolute) domains within the yz plane of MnWO$_4$, imaged with SIIG light from χ_{yzz} at 2.75 eV on an x-oriented sample ($d = 60~\mu$m). Similar to figure 7.2(b) a multiplicity of domains is obtained, indicated by black lines which coincide with the domain walls between the different absolute domains.

Subsequently, the order $\leftrightarrow$ disorder annealing cycle was applied which leads to the SHG image shown in figure 7.7(c). Most of the domain walls have vanished and only a small fraction of the sample remains in a domain state opposite to the rest of the sample. Apparently, the annealing procedure tends to drive the sample towards a single-domain state which was confirmed by repeating the experiment more than ten times.

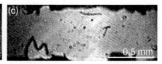

Figure 7.7: Response of the multiferroic domains to a temperature annealing cycle through the paramagnetic phase. (a) Schematic illustration of the applied annealing procedure. (b) Domain structure after initial zero-field cooling from room temperature to the multiferroic AF2 phase. (c) Domain structure after subsequent application of the annealing cycle. Black lines indicate the walls between absolute domains associated to order parameters of opposite orientation.

The tendency towards the formation of a single-domain state regarding the absolute domains is furhter supported by figures 7.6(a) and (c). Again, a single heating cycle through the paraelectric/-magnetic phase as sketched in figure 7.7(a) nearly induced a single-domain state.

Figure 7.7 leads to two conclusions. First, no memory effect is observed in the order ↔ disorder annealing cycle. Second, the observed behavior clearly separates the spin-spiral ferroelectric $MnWO_4$ from conventional ferroelectrics. While conventional ferroelectrics tend to form a multiplicity of domains for minimizing electric stray fields [48], the formation of a single-domain state seems energetically favorable in the case of $MnWO_4$ (Note that this applies to the absolute domains only as indicated by figures 7.6(b) and (d)).

In first place this behavior is quite surprising but can be explained taking into account that the system orders antiferromagnetically in the multiferroic phase. Ideal antiferromagnets tend to approach a single-domain state because they produce no macroscopic stray fields [174]. Moreover, the formation of magnetic domain walls costs additional energy and, therefore, is energetically not favorable.

Apparently, just like the domain structure itself, the "dynamic" domain topology is dominated by the magnetic aspect of the multiferroic hybrid domains.

7.4.2 Magnetic phase control by electric fields

Of special interest regarding the multiferroic domain state is the opportunity of a magnetic phase control by applying an alectric voltage, which is motivated in detail in the introductory section. Although the cross-correlation between the magnetic and electric properties of $MnWO_4$ has already been studied intensively by various integrating techniques [39, 96, 143, 144, 175], nearly nothing is known about the local electric-field response of the multiferroic domains in $MnWO_4$ or any other magnetically induced ferroelectric.

Therfore, we focus on the electric-field response of the absolute and the relative domains in this section.

Absolute and relative domain in electric fields

To investigate the electric-field response of the absolute and relative domains a static electric field $\mathcal{E}$ has been applied along the y axis. By measuring the integrated SHG intensity while sweeping the field, the relation between the magnetic order parameter η_2 (and P_y) and the electric field $\mathcal{E}$ can be probed. Moreover, spatially-resolved SHG measurements allow one to reveal the influence of $\mathcal{E}$ on the domain topology.

Figure 7.8 shows the response of absolute and relative domains to a static electric field $\mathcal{E}$ applied along the y axis. The presented SHG measurements are gained on an x-oriented sample which was rotated around its z axis by $\phi = 10°$. As discussed in detail in section 7.3, this configuration enables mapping of both types of domains that arise in the multiferroic phase of MnWO$_4$.

In figure 7.8(a) the measured (integrated) SHG intensity from $\chi^{\mathrm{ED}}_{yzz} + \chi_{\mathrm{cryst}}$ (2.75 eV) is shown as function of the applied electric field $\mathcal{E}||y$. The data is gained by cycling the electric field in the ± 3.5 kV/cm range.

Apparently, a butterfly loop is observed, clearly indicating the presence of a ferroelectric hysteresis. The butterfly shape is exclusively caused by the method and results from the interference of three contributions to the SHG intensity

$$I \propto |p(\mathcal{E}) \cdot \chi_{\mathrm{OP}} + \chi_{\mathrm{PARA}} \cdot e^{i\varphi} + \mathcal{E} \cdot \chi_{\mathrm{EFISH}} \cdot e^{i\varphi'}|^2 |\vec{E}(\omega)|^4 \tag{7.3}$$

with $-1 \leq p \leq +1$ as normalized ferroelectric polarization and $p = \pm 1$ as saturation values. The terms in equation (7.3) represent the order-parameter-related contribution[4] (χ_{OP}), the paramagnetic background contribution (χ_{PARA}), and a contribution scaling with the applied electric field $\mathcal{E}$ (χ_{EFISH}). The latter one is documented as "electric-field-induced second harmonic" (EFISH) in the literature [176,177]. The phase shift between these SHG contributions is parametrized by the angles φ and φ'.

By fitting equation (7.3) to the SHG data in figure 7.8(a) the actual hysteresis $p(\mathcal{E})$ is extracted and shown in figure 7.8(b). In agreement with dielectric measurements [39] the coercive field is about ± 2.8 kV/cm at $T = 10$ K.

Figures 7.8(c) and (d) show SHG images of the absolute domains obtained by zero-field cooling and by application of 3.5 kV/cm subsequent to zero-field cooling, respectively. As displayed in figure 7.8(c), a multi-domain state forms in the AF2 phase after zero-field cooling. Similar to figures 7.5(b), and 7.6(a, c) domains of opposite polarization P_y (and η_2) exhibit a different level of brightness. By applying an electric field $\mathcal{E}||y$ the sample is transformed into a single-domain state regarding the multiferroic hybrid (absolute) domains (figure 7.8(d)).

The effect of the poling procedure on the supplementary relative domains is revealed by figures 7.8(e) and (f). A comparison of figure 7.8(e) with figure 7.8(c) indicates that two relative domains, denoted by (1) and (2), are present within one of the absolute domains. However, figure 7.8(f) demonstrates that no relative domains are present after poling the sample. Thus, figures 7.8(e) and (f) evidence that the multiferroic domain state, including absolute and relative domains, is uniquely controlled by an electric field.

By application of an electric voltage the sample is transformed into a ferroelectric and magnetic single-domain state, meaning that no magnetic translation domains further subdivide the hybrid-multiferroic single-domain state. Domain walls between magnetic translation domains are "dragged along" as P_y (and η_2) reverses.

Figure 7.8(b) shows that repeated polarization reversal switches the multiferroic domain state repeatedly. The single-domain states encountered by this poling procedure are the same in each field cycle so that the SHG loop is always closed as seen in figure 7.8(b). Hence, corresponding to the hybrid nature of the multiferroic domains in MnWO$_4$, figures 7.8(a, b) can be interpreted as ferroelectric and also as antiferromagnetic hysteresis.

Note that the response of the split-order-parameter multiferroic BiFeO$_3$ to an electric field is pronouncedly different [178,179]. In BiFeO$_3$ the ferroelectric order arises independent

---

[4]Note that SHG from $\chi^{\mathrm{ED}}_{yzz}(+\chi_{\mathrm{cryst}})$ couples to the magnetic order parameter η_2 and, via equation (5.2), to P_y (see section 6.1.2).

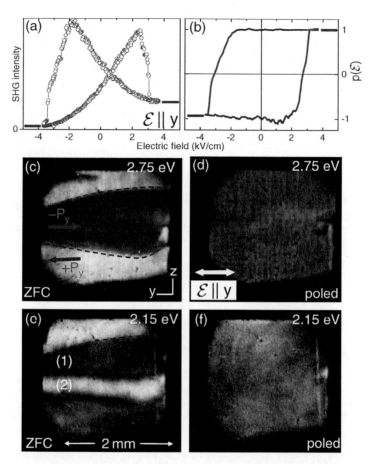

Figure 7.8: Control of the multiferroic domain state by electric fields. (a) Intensity of "multiferroic" SHG ($\chi_{yzz}^{\mathrm{ED}} + \chi_{\mathrm{cryst}}$ at 2.75 eV) in electric-field cycles of up to ± 3.5 kV/cm applied along the y axis. Bars denote the end points of the hysteresis. (b) Hysteresis derived from (a) on the basis of equation (7.3). Images: (c, d) Absolute domains imaged with SHG from $\chi_{yzz}^{\mathrm{ED}} + \chi_{\mathrm{cryst}}$ (10 K) taken after zero-field cooling (ZFC) (c) and after subsequent poling in an electric field of 3.5 kV/cm (d). Absolute and relative domains before (e) and after (f) poling, gained from $\chi_{yyy}^{\mathrm{ED}} + \chi_{\mathrm{cryst}}$ (2.15 eV)

of the magnetic order so that an electric field, though manipulating the magnetic domain population, does not set a unique magnetic single-domain state.

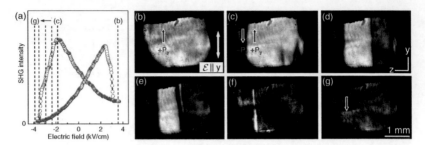

Figure 7.9: Local switching behavior of the multiferroic domains in electric fields $\mathcal{E}||y$. (a) SHG intensity in electric fields. At specific points the measurement has been interrupted to image the topology of the absolute domains in the yz plane of MnWO$_4$. (b – g) Spatially-resolved measurements of SHG from $\chi_{yzz} + \chi_{\mathrm{cryst}}$ at a photon energy of 2.75 eV. Apparently, the kinetics of the multiferroic domains is identical to the kinetics of conventional ferroelectric systems as detailed in the text.

Reversible switching

The previous sections revealed that electric-fields uniquely control the domain state of MnWO$_4$ (see figures 7.8 and 7.3). The application of an electric field allows to induce a single-domain state of the absolute as well as the relative domains. Thus, we now focus on the electric-field-control of multiferroic hybrid (absolute) domains only.

The domain structure in electric fields $\mathcal{E}$ in figures 7.3(b) and (c) already reflected the ferroelectric aspect of the multiferroic hybrid domains. Figure 7.9 now elucidates in detail the local switching behavior of these domains. SHG images are gained with light from $\chi_{yzz}^{\mathrm{ED}} + \chi_{\mathrm{cryst}}$ with $k||x$ (2.75 eV) while sweeping the electric field as indicated by the hysteresis in figure 7.9(a).

Surprisingly, the snap-shots obtained during the switching process exhibit exactly the same characteristics as known from conventional ferroelectrics, irrespective of the fact that the spontaneous polarization in MnWO$_4$ is strongly correlated to the magnetic properties. The typical stages in ferroelectric materials regarding the domain kinetics are [180]

1. Domain nucleation.

2. Forward growth of nucleation domains along the field direction (here $\mathcal{E}||y$).

3. Sideways domain growth via lateral domain wall movement in the direction normal to the polar y axis.

4. Coalescence of domains, tending to a single-domain state.

Obviously, all stages emerge in figure 7.9. Figure 7.9(c) reflects the transit growth of the multiferroic domains in the field direction, followed by a lateral movement of the domain walls in figure 7.9(d) until the single-domain state is finally reached in 7.9(g).

In conclusion, figure 7.9 proves that the kinetics of magnetically induced ferroelectric domains is equivalent to conventional ferroelectrics, though the mechanisms leading to the spontaneous polarization are fundamentally different.

The analogy can further be expanded to other qualities of a ferroelectric as the temperature dependence of the coercive field E_C which is the issue of figure 7.10(a) and 7.10(b).

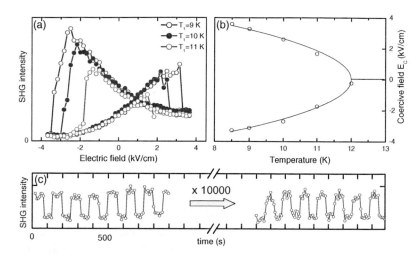

Figure 7.10: Performance of magnetically induced ferroelectricity under external electric fields. (a) Temperature dependence of the hysteresis associated to the ferroelectric order parameter. Presented curves correspond to the integrated SHG intensity related to χ_{yzz}, measured while sweeping the electric field. (b) Equivalent to conventional ferroelectrics, the coercive field extracted from (a) is proportional to P_y, leading to $E_C(T) \propto (T_2 - T)^{\frac{1}{2}}$ as indicated by the fitted curve (red). (c) Repeated switching of the multiferroic domain state. Even after a large number of electric-field cycles, no effects of fatigue are observed.

Figure 7.10(a) displays the hysteretic behavior of the ferroelectric order parameter measured at different temperatures in the multiferroic AF2 phase. Apparently, the coercive-field strength E_C decreases with increasing temperature and vanishes around $T_2 \approx 12$ K. Extracted values E_C are summarized in figure 7.10(b) showing the temperature dependence of the coercive field. For ferroelectric textbook materials, the coercive field is proportional to the spontaneous polarization $E_C \propto P$ [180].
According to section 5.1.3 this would imply

$$E_C \propto (T_2 - T)^{\frac{1}{2}} \qquad (7.4)$$

for MnWO$_4$ which is in striking agreement with the experimental observations shown by the fit (red curve) in figure 7.10(b).
Hence, the multiferroic domains exhibit all typical characteristics of a ferroelectric system. The improper character of the magnetically induced ferroelectricity is reflected by the smallness of the spontaneous polarization only. Moreover, figures 7.9 and 7.10 lead to the conclusion that the pseudo-proper behavior of the ferroelectric order parameter discussed in section 5.1.3 can be transfused to the associated domain structure acting exactly as conventional ferroelectric domains despite the strong correlation to the magnetic subsystem. The measurements presented in figures 7.9 and 7.10 reveal that magnetically induced ferroelectricity in principle provides the same qualities as usual ferroelectric materials.
To investigate also the "long-time" performance regarding the electric field control of the multiferroic domain state, the reversible switching in an alternating electric-field was mea-

sured. Figure 7.10(c) shows the integrated SHG intensity in alternating electric-fields of
± 3.5 kV/cm at $T = 8$ K. Discrete jumps in the SHG intensity indicate the switching
between multiferroic single-domain states corresponding to $\pm P_y$.

The data persented in 7.10(c) was gained by measuring the integrated SHG intensity while
reversing the electric field every 100 sec. After ensuring by this procedure that the sample
is transformed into a single-domain state by the applied electric field, the direction of $\mathcal{E}$
was changed more rapidly ($\delta t \approx 0.5$ sec). After about 10000 field cycles the SHG intensity
was measured again while continuing to reverse the field.

Even after 10000 cycles no changes in the performance are observable. The SHG yield related to $\pm P_y$ is unaffected and SHG images revealed that the whole sample still transforms
from one single-domain state into the other without a change of E_C.

Apparently, the repeated switching causes no major defects in the sample which may lead
to a local pinning of the multiferroic domain structure – no fatigue at all is obtained.

The switching without fatigue is quite remarkable and, consistent with figure 6.11, points
to an electronic origin of the ferroelectric polarization P_y. Since no ionic movements
accompany the switching process, defects in the crystal structure that are typically caused
by moving ions are avoided. Thus, the robustness of joint-order-parameter multiferroics as
$MnWO_4$ should be extraordinarily high which may allow for "fatigue-free switching".

The most interesting feature of a purely electronic polarization is its potential to further
reduce the switching-time limit of ferroelectric information bits.

In conventional ferroelectrics the lower limit for the switching time t_0 is given by the time
a domain wall needs to propagate from the top to the bottom electrode (or reversal) in a
sample with the thickness d:

$$t_0 = \frac{d}{\nu} \, . \tag{7.5}$$

Assuming that the upper limit of the domain wall velocity corresponds to the sound velocity $\nu_s = 4000$ m/s, the theoretical switching time t_0 is about 50 ps for a 200 nm thick
ferroelectric capacitor [181].

For a rough estimation of the switching velocity of a purely electronic polarization one may
consider the mobility of electrons described by the fermi-velocity which is in the order of
10^6 m/s ($\nu_{Cu} = 1.57 \cdot 10^6$ m/s [182]). Thus, according to equation (7.5), the fermi-velocity
as a limiting factor instead of the sound velocity would speed-up ferroelectric switching by
a factor of 100, leading to $t_0 < 1$ ps.

7.4.3 Local magnetoelectric memory effect

In section 7.4.1 it was shown that an order $\leftrightarrow$ disorder annealing cycle erases the multiferroic domain structure in favor of an almost single-domain state. The significantly
different domain structures imaged before and after the annealing procedure indicate that
absolutely no correlations exist between the two states. Moreover, no pinning effects are
observed.

The situation changes completely when an order $\leftrightarrow$ order annealing cycle from the AF2
phase to the AF1 phase and back to the multiferroic state is considered. This transition is
of particular interest, because of a magnetoelectric memory effect reported earlier [27,144]:
A single-domain state in the multiferroic phase is memorized in the nonpolar AF1 phase
and reemerges when reentering the AF2 phase.

However, all current data was just gained by integral techniques such as pyroelectric current
measurements, so that it is not known to what extent the memory effect applies to the

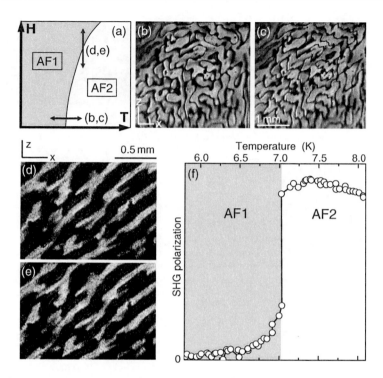

Figure 7.11: Response of the multiferroic domains to magnetic-field (H) and temperature (T) annealing cycles through the AF1 phase. (a) Sketch of the (H, T)-phase diagram with arrows indicating the respective annealing procedure. (b, d) Domain structure after initial zero-field cooling from room temperature to the multiferroic AF2 phase. (c, e) Domain structure after application of the respective annealing cycle. (f) Temperature dependence of the SHG signal (χ^{ED}_{yxx} component at 1.95 eV) in a temperature decreasing run. Nonzero SHG yield in the AF1 phase points to a residual polar contribution in the AF1 phase that may be the basis of the topological memory effect revealed in panels (b – e).

domain structure of a multiferroic multi-domain state.

This is investigated in figure 7.11, which compares the domain structure of the AF2 phase before (panels (b), (d)) and after (panels (c), (e)) the annealing cycle through the AF1 phase. All images were gained on y-oriented samples at $T = 10\ K$ with SHG from $\chi^{ED}_{x'z'z'}$ ($k||y$).

Figure 7.11(a) illustrates that the phase boundary between the AF2 and the AF1 phase can be crossed by magnetic-field or temperature tuning so that the memory effect can be investigated in dependence of either. Figures 7.11(b) and 7.11(c) reveal the effects of thermal annealing. Minor changes in the roughness of the domain walls and, for a small fraction of the walls, shifts in the order of 100 μm are observed but the general distribution of the domains is preserved. Annealing in a magnetic field along the y axis

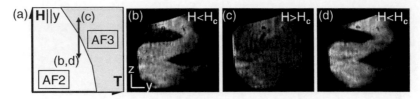

Figure 7.12: Domain-wall-related memory effect at the AF2 ↔ AF3 phase boundary. (a) The shape of the phase boundary between the AF3 and the AF2 phase allows to quench the multiferroic state by application of magnetic fields H parallel to the y axis of the crystal. (b) Multiferroic domain structure at $T = 12$ K after zero-field cooling. (c) A magnetic field $H_y > H_c$ quenches the multiferroic domain structure. (d) By returning to $H = 0$ T the initial domain structure reemerges.

reveals no changes at all in the domain structure[5]. Hence, not only a single-domain state is memorized in the non-chiral nonpolar AF1 phase – actually the entire topology of a multi-domain state is preserved.

The memory effect is thus much more rigid than established up to now. It allows one to hide the multiferroic domain state by quenching it in the transition to the AF1 phase and reconstructing the domain structure (near-) identically upon return to the AF2 phase.

Figure 7.11 immediately raises the question for the origin of such a memory effect. Pinning of the domain structure by major structural defects can be excluded. This mechanism would also preserve the domain structure in the order ↔ disorder transition to the paramagnetic phase contrary to what figure 7.7 shows.

For revealing its origin, the temperature dependence of multiferroic SHG from χ^{ED}_{yxx} was measured across the AF2 → AF1 transition (see section 6.1.2). As shown in Fig 7.11(f), the SHG contribution does not vanish abruptly at the phase boundary to the AF1 state. Instead, it diminishes gradually across an interval of > 1 K after an initial step-like decrease. The remanent SHG signal reveals a coexistence of the AF1 and AF2 phase within a broad temperature region, enabled by the first-order nature of the transition. Hence, residual nuclei of the AF2 phase explain the "polar memory" of $MnWO_4$ [27]. Such a distribution of the nuclei must be homogeneous and dense in order to explain pinning of the entire topology of a multi-domain state.

The manifestation of the ferroelectric polarization in the temperature-dependent measurement shown in figure 7.11(f) is quite remarkable, because it was not observed by pyroelectric measurements. This can be attributed to the extraordinary sensitivity of SHG to nanoscopic inclusions [183].

Another type of memory effect is observed in $MnWO_4$ when the multiferroic AF2 phase is suppressed by magnetic fields. As schematically displayed in figure 7.12(a), magnetic fields $H \| y$ destabilize the AF2 phase while broadening the stability range of the AF3 phase towards lower temperatures. Thus, it is possible to quench the multiferroic phase by a magnetic field $H_y > H_c$ applied along the y axis of the crystal. Here, H_c denotes the critical magnetic field associated to the phase boundary between the AF2 and AF3 phase in the (H,T)-phase diagram (see also figure 5.3).

Figure 7.12 demonstrates that the multiferroic domain state in figure 7.12(b) is quenched

---

[5]Due to a small misalignment of the sample, a crystallographic contribution admixes with SHG from $\chi^{ED}_{x'z'z'}$ leading to the contrasts are observed in figures 7.11(d) and 7.11(e).

by a magnetic field $H_y > 4.3$ T at $T = 12$ K (figure 7.12(c)). Surprisingly, the multiferroic domain topology reemerges in figure 7.12(d) when the magnetic field is removed.

Apparently, the suppression of the multiferroic domain state not necessarily erases the complete information about the former domain topology. As indicated by figure 7.12 the initial structure is not irretrievably destroyed when the AF3 phase is induced.

Note that the memory effect obtained at the AF2 ↔ AF3 transition is qualitatively different from the magnetoelectric memory effect in figure 7.11:

Ferroelectric nuclei are considered responsible for the memorized ferroelectric domain state in the AF1 phase which is supported by figure 7.11(f) [27]. However, the existence of such ferroelectric nuclei is restricted to first-order phase transitions as observed at T_3.

In contrast, the phase transition at T_2 is a second-order phase transition, showing no hysteretic behavior. Thus, no region of coexisting phases AF2 and AF3 is expected that might explain a conservation of the multiferroic domain structure in figure 7.12. Moreover, the observed memory effect does not appear in pyroelectric measurements when poled ferroelectric single-domain states are investigated [27].

This observation suggests that the low symmetry at the multiferroic domain walls may play an essential role regarding the local memory effect observed in the AF3 phase. Since no domain walls are present in the case of a poled sample, a single-domain state is simply erased by quenching the multiferroic AF2 phase in magnetic fields. As a consequence a multi-domain state emerges as the magnetic field is removed again, which cannot be resolved by pyroelectric measurements.

Note that a mechanism associated to domain walls as proposed here is different from the one we proposed in PRL **102**, 107202 (2009). In the aforementioned publication we attributed the obtained memory effect to the periodic modulation and related wave vectors of the spin-density wave in phase AF3. However, the progressing experiments and a better understanding of the complex magnetic order in $MnWO_4$ revelead that the mechanism suggested earlier cannot be the origin of the measured memory effect.

In conclusion, a domain-wall-related memory effect in the AF3 phase would explain the observation presented in figure 7.12 as well as the absence of a memory effect in the case of pyroelectric measurements [27]. However, to consolidate the proposed role of domain walls a theoretical model is highly desirable to explain the magnetic-field driven conceiling of the multiferroic domain state.

Chapter 8

Conclusions and Outlook

In the framework of this thesis, the magnetoelectric interactions in magnetically-induced ferroelectrics have been investigated. In these materials magnetic long-range order breaks the spatial inversion symmetry and induces a spontaneous polarization, leading to an intrinsically strong and robust coupling.

Optical second harmonic generation (SHG) is applied and proven explicitly feasible for studying these complex and challenging materials.

The systematic analysis of SHG contributions reveals that the symmetry-breaking order parameters of the multiferroic state are inseparably entangled in magnetically-induced ferroelectrics, reflecting the intrinsically strong coupling of magnetism and ferroelectricity in these compounds. Most remarkably, the entanglement also manifests in the local domain structure, leading to the formation of a novel type of domains. These domains are introduced as "hybrid-multiferroic domains", indicating their bilateral (magnetic and electric) nature.

The observation of the rigid one-to-one correlation of magnetic and electric degrees of freedom on the level of domains evidences that the unique properties of magnetically-induced ferroelectrics could indeed be exploited in multifunctional devices. Moreover, introductory field-dependent measurements of the local domain structure unambiguously show that an electric voltage uniquely controls the magnetic domain state. Although the obtained spontaneous polarization is typically small, it its the robustness of the inherent magnetoelectric coupling that renders magnetically-induced ferroelectrics interesting with respect to long-term applications.

An interesting topic for future experiments would be the dynamic aspect of the presented magnetic phase control. Pump-probe SHG measurements using the advantage of femtosecond laser systems are explicitly valuable in this context, because they allow to study the dynamical properties of the magnetic *and* the electric subsystem with a resolution in the femtosecond regime.

Besides providing information about the dynamical limits of a magnetic phase control by electric fields in magnetically-induced ferroelectrics, femtosecond-pump-probe experiments may ultimately clearifiy whether electronic or ionic contributions cause the spontaneous polarization in magnetically-induced ferroelectrics: Since charge redistributions and ionic displacements occur on different time scales [184], they should be distinguishable in the time domain.

It would also be interesting to investigate the dynamics of multiferroic hybrid domains in systems with conical spin-spiral, because these materials display a macroscopic magnetization. Since magnetization and polarization are manifestations of the same multiferroic hybrid domain state, an electric field will always act on both. Rapid magnetization reversal with ultrashort electric-field pulses may thus become feasible.

However, although a vastly increasing variety of multiferroic materials that host rather strong magnetoelectric interactions are at our disposal for multiferroic domain control, multiferroics have not entered the mainstream products yet. This is mainly due to the fact that "strong coupling effects" are synonymous to "low temperature" in most of the studied multiferroics.

Thus, an important question is: What will the future of multiferroics hold? Of special interest in this context are thin-film heterostructures and multilayers which provide a pathway to create multiferroicity in an artificial way. In these structures, the two-phase coexistence that constitutes multiferroic behavior is often realized at interfaces.

The remarkable advances in the growth of thin films and the ability to synthesize complex superlattices of different materials even allow for "designing" novel interface effects between electronic spins, charge, and orbitals. By this approach, structures can be created that intrinsically lack inversion symmetry, although its constituents are centrosymmetric. In particular, spatial- and time-reversal symmetry are violated at the interface of two magnetically ordered materials, a prerequisite for displaying a linear magnetoelectric effect.

This field of material engineering is still in the beginning and strongly benefits from the ongoing progress regarding the applied deposition techniques. Thus, sooner or later the combination of our better understanding of the fundamentals in multiferroic materials and improved sophisticated growth techniques will allow for developing multiferroic devices that display an applicable magnetoelectric effect at room-temperature.

Appendix A

Phase matching in MnWO$_4$

The investigation of the multiferroicity in MnWO$_4$ revealed different interesting optical properties beyond the previously discussed phenomena. Since these properties are not relevant for the analysis of the symmetry-breaking order parameters or the domain topology of the system, they are presented separately in this appendix.

The SHG spectra shown in section 6.1 indicate that the efficiency of the SHG process is extraordinarily high regarding SHG from χ_{yxx}^{ED}. A comparison of the peak intensity obtained at 1.95 eV exceeds SHG from χ_{yzz}^{ED} or $\chi_{x'z'z'}^{\mathrm{ED}}$ by a factor of 100 (see figure 6.2). Moreover, the peak position (1.95 eV) deviates from the transition energy corresponding to the associated d–d transition $^6A_{1g} \rightarrow {}^4T_{1g}$ (2.15 eV).

As already mentioned, both features are hallmarks of so-called phase-matching which is discussed in detail in the following.

The strongest SHG signals typically appear when the refractive index of a material is equal for the incident light at frequency ω and SHG light at 2ω, i.e. $n(\omega) = n(2\omega)$. In this case the generated wave is able to extract energy most efficiently from the incident waves [126]. This is expressed by

$$\Delta k = 2k(\omega) - k(2\omega) = 0 \ , \tag{A.1}$$

known as *perfect phase matching condition* with $k(\omega) = \frac{\omega}{c_0} n(\omega)$. For $\Delta k = 0$ the conversion efficiency is at its maximum and the SHG yield increases with the length ℓ of the investigated crystal according to figure A.1(a) [185].

Reversley, a wave vector mismatch $\Delta k \neq 0$ lowers the conversion efficiency of the SHG

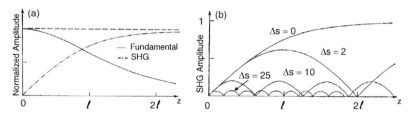

Figure A.1: (a) Normalized SHG intensity in the case of perfect phase matching $\Delta k = 0$ (Graphs are taken from reference [185]). It can be seen that the SHG intensity increases with the thickness ℓ of the sample. (b) A phase mismatch $\Delta s = \Delta k \ell \neq 0$ between the fundamental light field and the SHG signal leads to a drastic decrease of the SHG intensity.

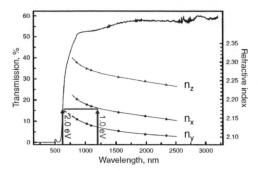

Figure A.2: Transmission (black curve) and refractive indexes for x-, y-, and z-polarized light of MnWO$_4$ taken from reference [137]. The blue and the red arrow indicate that the phase matching condition (A.1) is fulfilled for y-polarized SHG induced by x-polarized incident light fields.

process as depicted in figure A.1(b). Here, $\Delta s = \Delta k \ell$ is a normalized measure for the mismatch of the light fields. A value $\Delta s = 0$ indicates perfect phase matching, so that the corresponding curve in figure A.1(b) is the same as shown in figure A.1(a).

Figure A.1(b) reveals that the SHG intensity drastically decreases for a mismatch ($\Delta s \neq 0$) of the participating light fields. As shown in detail in reference [126], this can also be expressed by

$$I^{\mathrm{SHG}} = I^{\max} \frac{\sin^2(\Delta k \ell)}{(\Delta k \ell / 2)^2} \; , \qquad (A.2)$$

describing the SHG intensity I^{SHG} as function of the mismatch, parametrized by $\Delta k \ell$.

Figures A.1(a),(b), and equation (A.2) point out that the SHG yield in a birefringent crystal is influenced by two parameters, i.e. the wavelength of the incident light field which determines the mismatch Δk and the effective path length ℓ through the sample [186].

We now come to the investigated compound MnWO$_4$. The refractive indexes of MnWO$_4$ for x-, y-, and z-polarized light as function of the wavelength are shown in figure A.2 which is taken from reference [137].

Figure A.2 explicitly shows that the phase matching condition A.1 is fulfilled for y-polarized SHG (≈ 2.00 eV $= 620$ nm) obtained from x-polarized incident light fields (≈ 1.00 eV $= 1240$ nm). Thus, figure A.2 allows to understand the pronounced nonlinear response considering SHG from χ^{ED}_{yxx} in the SHG spectrum presented in figure 6.2(b).

The phase matching also explains the shift in energy with respect to the actual transition energy of the related d–d excitation. Although explicitly strong SHG signals are typically expected for a resonant excitation (see chapter 4) phase matching seems to be the dominant effect in the present case so that the maximum intensity is obtained at a slightly lower energy. This leads to the conclusion that phase matching generally plays an important role for the shape of measured SHG spectra in MnWO$_4$.

The astonishing complexity of phase matching in MnWO$_4$ is revealed by figure A.3(a), showing the SHG intensity related to χ^{ED}_{yxx} and/or χ^{ED}_{yzz} for various directions of the incident light field. Figure A.3(a) shows the spectral dependence of y-polarized SHG gained from

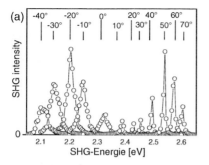 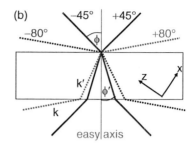

Figure A.3: (a) Spectral dependence of y-polarized SHG obtained with x, z-polarized light (8 K). The obtained peak position strongly depends on the direction of the incident light field which is parametrized by ϕ. For $-\phi$ ($k' \rightarrow z$) the peak shifts to lower energies, whereas a rotation by $+\phi$ ($k' \rightarrow x$) has the opposite effect. (b) Schematic illustration of the experimental setup. Light fields are incident in the xz plane of MnWO$_4$ while ϕ describes the angle between k and the easy axis of the system. The angle ϕ is changed by a rotation of the sample around its y axis.

x, z-polarized light[1] for various propagation directions k of the fundamental beam, here parametrized by the angle ϕ.

The measurements summarized in figure A.3(a) are realized by using a sample with polished surfaces normal to the easy axis of the system. Figure A.3(b) schematically illustrates the geometry of the experiment: While initial light fields were chosen to propagate in the xz plane of MnWO$_4$, ϕ denotes the angle between their propagation direction k and the easy axis. Note that the propagation direction k' in the sample significantly deviates from k due to refraction at the sample surface.

The relation between k and k' is sketched in figure A.3(b) for chosen angles ϕ in between $-80°$ and $+80°$.

However, for normal incidence $\phi = 0°$ ($k||$(easyaxis)) the SHG signal in figure A.3(a) peaks around 2.32 eV. When the sample is rotated around its y axis by $-\phi$, the direction of k' approaches the direction of the z axis of the sample and the peak shifts to lower energies. This trend is constistent with figure 6.2(b), where the peak is observed around 2.00 eV for $k||z$.

In contrast, a rotation in the opposite direction by $+\phi$ results in a shift of the peak to higher energies. Apparently, the peak position obtained with $k||x$ (2.75 eV) is continuously approached.

The correlation between the angle ϕ and the peak position in figure A.3(a) can be understood on the basis of the refractive indexes given by figure A.2. The blue and the red arrow indicate that the phase matching condition (A.1) is satisfied for y-polarized SHG from x-polarized light. As the sample is rotated around y, the refractive index relevant for the incident light fields increases ($n_x < n_z$) whereas the refractive index for the SHG n_y remains constant. As a consequence the phase matching condition is satisfied for a smaller

[1]Here, the direction of incident light fields does not coincide with one of the principle crystal axes. Thus, light fields polarized perpendicular to the y axis have nonzero components E_x and E_z while their magnitude depends on the angle ϕ which determines the projection of $\mathbf{E}$ on x and z. See figure A.3(b).

wavelength, i.e. at a higher energy.

This is exactly what we see in figure A.3(a). An angle of $\phi = -40°$ means that the direction of the incident light field is rather close to the z axis (see figure A.3(b)). By the rotation $-\phi \rightarrow +\phi$ the angle between k' and the z axis increases and so does the refractive index $(n_x \rightarrow n_z)$. Therefore, the peak that indicates phase matching shifts to higher energies.

In conclusion, figure A.3 shows that the spectral dependence of y-polarized SHG obtained from x- and z-polarized incident light is predominantly determined by the propagation direction of the fundamental beam. This observation emphasizes the general importance of phase matching and its impact on the shape of polarization-dependent SHG spectra in MnWO$_4$.

Another feature of figure A.3(a) is quite remarkable. A comparison with figure 6.1(a) reveals that the envelope of the data in figure A.3(a) is determined by the linear absorption of MnWO$_4$. First of all, a maximum of the SHG yield occurs around 2.2 eV corresponding the maximum at 2.15 eV in α. The offset in energy simply occurs due to an increase of the effective length ℓ of the sample for rather large angles ϕ so that absorption becomes more important.

Second, both envelope and linear absorption exhibit a minimum at 2.45 eV which is followed by a significant increase for higher photon energies. Only above ≈ 2.55 eV the two curves show a different spectral shape. This follows from the drastic increase of the absorption $(\phi \geq 50°)$, leading to the observed suppression of the SHG signal in figure A.3(a).

Bibliography

[1] Tokura, Y. & Nagaosa, N. Orbital physics in transition-metal oxides. *Science* **288**, 462–468 (2000).

[2] Dagotto, E. Complexity in strongly correlated electronic systems. *Science* **309**, 257–262 (2005).

[3] Bednorz, J. G. & Müller, K. A. Possible high T_c superconductivity in the Ba-La-Cu-O system. *Z. Phys. B: Condens. Matter* **64**, 189–193 (1986).

[4] Binasch, G., Grünberg, P., Saurenbach, F. & Zinn, W. Enhanced magnetoresistance in layered magnetic structures with antiferromagnetic interlayer exchange. *Phys. Rev. B* **39**, 4828–4830 (1989).

[5] Baibich, M. N. *et al.* Giant magnetoresistance of (001)Fe/(001)Cr magnetic sublattices. *Phys. Rev. Lett.* **61**, 2472–2475 (1988).

[6] Helmolt, R. v., Wecker, J., Holzapfel, B., Schultz, L. & Samwer, K. Giant negative magnetoresistance in perovskitelike $La_{2/3}Ba_{1/3}MnO_x$ ferromagnetic films. *Phys. Rev. Lett.* **71**, 2331–2333 (1993).

[7] Chahara, K., Ohno, T., Kasai, M. & Kozono, Y. Magnetoresistance in magnetic manganese oxide with intrinsic antiferromagnetic spin structure. *Appl. Phys. Lett.* **63**, 1990 (1993).

[8] Kimura, T. *et al.* Magnetic control of ferroelectric polarization. *Nature* **426**, 55–58 (2003).

[9] Hur, N. *et al.* Electric polarization and memory in a multiferroic material induced by magnetic fields. *Nature* **429**, 392–395 (2004).

[10] Lottermoser, T. *et al.* Magnetic phase control by an electric field. *Nature* **430**, 541–545 (2004).

[11] Goto, T., Kimura, T., Lawes, G., Ramirez, A. P. & Tokura, Y. Ferroelectricity and giant magnetocapacitance in perovskite rare-earth manganites. *Phys. Rev. Lett.* **92**, 257201 (2004).

[12] Lawes, G. *et al.* Magnetically driven ferroelectric order in $Ni_3V_2O_8$. *Phys. Rev. Lett.* **95**, 087205 (2005).

[13] Yamasaki, Y. *et al.* Magnetic reversal of the ferroelectric polarization in a multiferroic spinel oxide. *Phys. Rev. Lett.* **96**, 207204 (2006).

[14] Senff, D. *et al.* Magnetic excitations in multiferroic TbMnO$_3$: Evidence for a hybridized soft mode. *Phys. Rev. Lett.* **98**, 137206 (2007).

[15] Ishiwata, S., Taguchi, Y., Murakawa, H., Onose, Y. & Tokura, Y. Low-magnetic-field control of electric polarization vector in a helimagnet. *Science* **319**, 1643–1646 (2008).

[16] Zvezdin, A. K., Logginov, A. S., Meshkov, G. A. & Pyatakov, A. P. Multiferroics: Promising materials for microelectronics, spintronics, and sensor technique. *Bull. Rus. Acad. Sci., Phys.* **71**, 1561–1562 (2007).

[17] Dawber, M., Rabe, K. M. & Scott, J. F. Physics of thin-film ferroelectric oxides. *Rev. Mod. Phys.* **77**, 1083 – 1130 (2005).

[18] Chen, X., Hochstrat, A., Borisov, P. & Kleemann, W. Magneto-electric exchange bias systems in spintronics. *Appl. Phys. Lett.* **89**, 202508 (2006).

[19] Bibes, M. & Barthélémy, A. Multiferroics: Towards a magnetoelectric memory. *Nat. Mater.* **7**, 425–426 (2008).

[20] Béa, H., Gajek, M., Bibes, M. & Barthélémy. Spintronic with multiferroics. *J. Phys.: Condens. Matter* **20**, 434211 (2008).

[21] Ramesh, R. & Schlom, D. G. Whither oxides electronic? *MRS Bull.* **33**, 1006–1011 (2008).

[22] Zhao, T. *et al.* Electrical control of antiferromagnetic domains in multiferroic BiFeO$_3$ films at room temperature. *Nat. Mater.* **5**, 823–829 (2006).

[23] Choi, T., Lee, S., Choi, Y. J., Kiryukhin, V. & Cheong, S.-W. Switchable ferroelectric diode and photovoltaic effects in bifeo$_3$. *Science* **324**, 63–66 (2009).

[24] Aliouane, N. *et al.* Flop of electric polarization driven by the flop of the mn spin cycloid in multiferroic TbMnO$_3$. *Phys. Rev. Lett.* **102**, 207205 (2009).

[25] Bodenthin, Y. *et al.* Manipulating the magnetic structure with electric fields in multiferroic ErMn$_2$O$_5$. *Phys. Rev. Lett.* **100**, 027201 (2008).

[26] Choi, Y. J. *et al.* Thermally or magnetically induced polarization reversal in the multiferroic CoCr$_2$O$_4$. *Phys. Rev. Lett.* **102**, 067601 (2009).

[27] Taniguchi, K., Abe, N., Ohtani, S. & Arima, T. Magnetoelectric memory effect of the nonpolar phase with collinear spin structure in multiferroic MnWO$_4$. *Phys. Rev. Lett.* **102**, 147201 (2009).

[28] Eerenstein, W., Mathur, N. D. & Scott, J. F. Multiferroic and magnetoelectric materials. *Nature* **442**, 759–765 (2006).

[29] Cheong, S.-W. & Mostovoy, M. Multiferroics: a magnetic twist for ferroelectricity. *Nat. Mater.* **6**, 13–20 (2007).

[30] Katsura, H., Nagaosa, N. & Balatsky, A. V. Spin current and magnetoelectric effect in noncollinear magnets. *Phys. Rev. Lett.* **95**, 057205 (2005).

[31] Mostovoy, M. Ferroelectricity in spiral magnets. *Phys. Rev. Lett.* **96**, 067601 (2006).

[32] Kimura, T., Sekio, Y., Nakamura, H., Siegrist, T. & Ramirez, A. P. Cupric oxide as an induced-multiferroic with high-T_C. *Nat. Mater.* **7**, 291–294 (2008).

[33] Spaldin, N. A. & Fiebig, M. The renaissance of magnetoelectric multiferroics. *Science* **309**, 391–392 (2005).

[34] Khomskii, D. I. Multiferroics: different ways to combine magnetism and ferroelectricity. *J. Magn. Magn. Mater.* **306**, 1–8 (2006).

[35] Khomskii, D. Classifying multiferroics: Mechanisms and effects. *Physics* **2**, 20 (2009).

[36] Kimura, T. Spiral magnets as magnetoelectrics. *Annu. Rev. Mater. Res.* **37**, 387–413 (2007).

[37] Kimura, Y., T. Tokura. Magnetoelectric phase control in a magnetic system showing cycloidal/conical spin order. *J. Phys.: Condens. Matter* **20**, 434204 (2008).

[38] Loidl, A., Loehneysen, H. v. & Kalvius, G. M. Multiferroics. *J. Phys.: Condens. Matter* **20**, 430301 (2008).

[39] Kundys, B., Simon, C. & Martin, C. Effect of magnetic field and temperature on the ferroelectric loop in MnWO$_4$. *Phys. Rev. B* **77**, 172402 (2008).

[40] Radaelli, P. G. *et al.* Electric field switching of antiferromagnetic domains in YMn$_2$O$_5$: A probe of the multiferroic mechanism. *Phys. Rev. Lett.* **101**, 067205 (2008).

[41] Cabrera, I. *et al.* Coupled magnetic and ferroelectric domains in multiferroic Ni$_3$V$_2$O$_8$. *Phys. Rev. Lett.* **103**, 087201 (2009).

[42] Wadhawan, V. K. Towards a rigorous definition of ferroic phase transitions. *Phase Trans.* **64**, 165–177 (1998).

[43] Guymont, M. Symmetry analysis of structural transitions between phases not necessarily group-subgroup related. *Phys. Rev. B* **24**, 2647–2655 (1980).

[44] Liboff, R. L. *Primer for Point and Space Groups* (Springer-Verlag, New York, 2004).

[45] Tolédano, P. & Dmitriev, V. *Reconstructive Phase Transitions: In Crystals and Quasicrystals* (World Scientific, 1996).

[46] Tolédano, J. C. & Tolédano, P. *The Landau theory of phase transitions* (World Scientific, 1987).

[47] Birss, R. R. *Symmetry and Magnetism* (North-Holland Publishing Company, 1964).

[48] Wadhawan, V. K. *Introduction to ferroic materials* (Gordon and Breach Science Publishers, 2000).

[49] Bishop, D. M. *Group theory and chemistry* (Dover Publications, 1993).

[50] van Smaalen, S. *Incommensurate Crystallography* (Oxford: Oxford University Press, 2007).

[51] De Wolff, P. M. The pseudo-symmetry of modulated crystal structures. *Acta Crystallogr., Sect. A* **30**, 777–785 (1974).

[52] De Wolff, P. M. Symmetry operations for displacively modulated structures. *Acta Crystallogr., Sect. A* **33**, 493–497 (1977).

[53] Janssen, T. & Janner, A. Incommensurability in crystals. *Adv. Phys.* **36**, 519–624 (1987).

[54] Wagner, M. *Gruppentheoretische Methoden in der Physik* (Vieweg, 1998).

[55] Dresselhaus, M. S., Dresselhaus, G. & Jorio, A. *Group Theory: Application to the Physics of Condensed Matter* (Springer, Berlin, 2008).

[56] Aroyo, M. I. & Wondratschek, H. *International Tables for Crystallography, Volume B: Reciprocal Space, Band 2* (Springer, Berlin, 2008).

[57] Ashcroft, N. W. & Mermin, D. N. *Festkörperphysik* (Oldenbourg, 2007).

[58] Chaikin, P. M. & Lubensky, P. C. *Principles of condensed matter physics* (Cambridge University Press, 2000).

[59] Kovalev, O. V. *The irreducible representation of space groups* (Gordon and Breach, 1965).

[60] Aroyo, M. I., Kirov, A., Capillas, C., Perez-Mato, J. M. & Wondratschek, H. Bilbao crystallographic server ii: Representations of crystallographic point groups and space groups. *Acta Crystallogr., Sect. A* **62**, 115–128 (2006).

[61] Duan, F. & Guojun, J. *Introduction to Condensed Matter Physics, Volume 1* (World Scientific Publishing, 2005).

[62] Schmid, H. Multi-ferroic magnetoelectrics. *Ferroelectrics* **162**, 317–338 (1994).

[63] van Aken, B. B., Rivera, J.-P., Schmid, H. & Fiebig, M. Observation of ferrotoroidic domains. *Nature* **449**, 702–705 (2007).

[64] Aizu, K. Possible species of "ferroelastic" crystals and of simultaneously ferroelectric and ferroelastic crystals. *J. Phys. Soc. Jpn.* **27**, 387–396 (1969).

[65] Aizu, K. Possible species of ferromagnetic, ferroelectric, and ferroelastic crystals. *Phys. Rev. B* **2**, 754–772 (1970).

[66] Zheng, H. *et al.* Multiferroic $BaTiO_3$-$CoFe_2O_4$ nanostructures. *Science* **303**, 661–663 (2004).

[67] Agyei, A. K. & Birman, J. L. On the linear magnetoelectric effect. *J. Phys.: Condens. Matter* **2**, 3007–3020 (1990).

[68] Rivera, J.-P. On definitions, units, measurements, tensor forms of the linear magnetoelectric effect and on a new dynamic method applied to cr-cl boracite. *Ferroelectrics* **161**, 165 (1994).

[69] Fiebig, M. Revival of the magnetoelectric effect. *J. Phys. D: Appl. Phys.* **38**, R123–R152 (2005).

[70] Weiglhofer, W. S. & Lakhtakia, A. *Introduction to Complex Mediums for Optics and Electromagnetics* (Spie Press Monograph, 2003).

[71] O'Dell, T. H. *The Electrodynamics of Continuous Media* (North-Holland, Amsterdam, The Netherlands, 1970).

[72] Fiebig, M., Lottermoser, T., Fröhlich, D., Goltsev, A. V. & Pisarev, R. V. Observation of coupled magnetic and electric domains. *Nature* **419**, 818–820 (2002).

[73] Astrov, D. N. The magnetoelectric effect in antiferromagnetics. *Sov. Phys. JETP* **11**, 708 (1960).

[74] Astrov, D. N. Magnetoelectric effect in chromium oxide. *Sov. Phys. JETP* **13**, 729 (1961).

[75] www.chem.qmul.ac.uk.

[76] Ederer, C. & Spaldin, N. A. Electric-field switchable magnets: The case of $BaNiF_4$. *Phys. Rev. B* **74**, 020401(R) (2006).

[77] Kenzelmann, M. *et al.* Magnetic inversion symmetry breaking and ferroelectricity in $TbMnO_3$. *Phys. Rev. Lett.* **95**, 087206 (2005).

[78] Hill, N. A. Why are there do few magnetic ferroelectrics. *J. Phys. Chem. B* **104**, 6694–6709 (2000).

[79] Teague, J. R., Gerson, R. & James, W. J. Dielectric hysteresis in single crystal $BiFeO_3$. *Solid State Commun.* **8**, 1073–1074 (1970).

[80] Kiselev, S. V., Ozerov, R. P. & Zhdanov, G. S. Detection of magnetic order in ferroelectric $BiFeO_3$ by neutron diffraction. *Sov. Phys. Dokl.* **7**, 742–744 (1963).

[81] Moreira dos Antos, A., Parashar, A. R., S. andRaju, Zhao, Y. S., Cheetham, A. K. & Rao, C. N. R. Evidence for the likely occurence of magnetoferroelectricity in the simple perovskite, $BiMnO_3$. *Solid State Commun.* **122**, 49–52 (2002).

[82] Jonker, G. H. & Van Santen, J. H. Ferromagnetic compounds of manganese with perovskite structure. *Physica* **16**, 337–349 (1950).

[83] Seshadri, R. & Hill, N. A. Visualizing the role of bi 6s lone pairs in the off-center distortion in ferromagnetic $BiMnO_3$. *Chem. Mater.* **13**, 2892–2899 (2001).

[84] van Aken, B. B., Palstra, T. T. M., Filippetti, A. & Spaldin, N. A. The origin of ferroelectricity in magnetoelectric $YMnO_3$. *Nat. Mater.* **3**, 164–170 (2004).

[85] Lonkai, T. *et al.* Development of the high-temperature phase of hexagonal manganites. *Phys. Rev. B* **69**, 134108 (2004).

[86] van den Brink, J. & Khomskii, D. I. Multiferroicity due to charge ordering. *J. Phys.: Condens. Matter* **20**, 434217 (2008).

[87] Ikeda, N. *et al.* Charge frustration and dielectric dispersion in $LuFe_2O_4$. *J. Phys. Soc. Jpn.* **69**, 1526–1532 (2000).

[88] Ikeda, N. *et al.* Ferroelectricity from iron valence ordering in the charge-frustrated system LuFe$_2$O$_4$. *Nature* **436**, 1136–1138 (2005).

[89] Wen, J., Xu, G., Gu, G. & Shapiro, S. M. Magnetic field control of charge structures in the magnetically disordered phase of the multiferroic LuFe$_2$O$_4$. *Phys. Rev. B* **80**, 020403 (2009).

[90] Li, C.-H. *et al.* Electrical control of magnetization in charge-ordered multiferroic LuFe$_2$O$_4$. *Phys. Rev. B* **79**, 184430 (2009).

[91] Picozzi, S. & Ederer, C. First principles studies of multiferroic materials. *Psi-k network (www.psi-k.org)* **8**, 35–71 (2009).

[92] Tolédano, P., Schranz, W. & Krexner, G. Induced ferroelectric phases in TbMn$_2$O$_5$. *Phys. Rev. B* **79**, 144102 (2009).

[93] Tolédano, P. Pseudo-proper ferroelectricity and magnetoelectric effects in TbMnO$_3$. *Phys. Rev. B* **79**, 094416 (2009).

[94] Newnham, R. E., Kramer, J. J., Schulze, W. A. & Cross, L. E. Magnetoferroelectricity in Cr$_2$BeO$_3$. *J. Appl. Phys.* **49**, 6088–6091 (1978).

[95] Picozzi, S., Yamauchi, K., Sanyal, B., Sergienko, I. A. & Dagotto, E. Dual nature of improper ferroelectricity in a magnetoelectric multiferroic. *Phys. Rev. Lett.* **99**, 227201 (2007).

[96] Taniguchi, K., Abe, N., Takenobu, T., Iwasa, Y. & Arima, T. Ferroelectric polarization flop in a frustrated magnet MnWO$_4$ induced by a magnetic field. *Phys. Rev. Lett.* **97**, 097203 (2006).

[97] Heisenberg, W. Mehrkörperproblem und Resonanz in der Quantenmechanik. *Zeitschrift für Physik* **38**, 411–426 (1926).

[98] Schwabl, F. *Statistische Mechanik* (Springer, Berlin, 2000).

[99] Goodenough, J. B. *Magnetism and the Chemical Bond* (John Wiley and Sons, New York-London, 1963).

[100] Kimura, H. *et al.* Spiral spin structure in the commensurate magnetic phase of multiferroic RMn$_2$O$_5$. *J. Phys. Soc. Jpn.* **76**, 074706 (2007).

[101] Chapon, L. C. *et al.* Structural anomalies and multiferroic behavior in magnetically frustrated TbMn$_2$O5. *Phys. Rev. Lett.* **93**, 177402 (2004).

[102] Moriya, T. Anisotropic superexchange interaction and weak ferromagnetism. *Phys. Rev.* **120**, 91–98 (1960).

[103] Sergienko, I. A. & Dagotto, E. Role of the Dzyaloshinskii-Moriya interaction in multiferroic perovskites. *Phys. Rev. B* **73**, 094434 (2006).

[104] Dzyaloshinskii, I. E. On the magneto-electrical effects in antiferromagnets. *Sov. Phys. JETP* **10**, 628–629 (1959).

[105] Dzyaloshinskii, I. E. Theory of helical structures in antiferromagnets i: Nonmetals. *Sov. Phys. JETP* **19**, 960–971 (1964).

[106] Bruno, P. & Dugaev, V. K. Equilibrium spin currents and the magnetoelectric effect in magnetic nanostructures. *Phys. Rev. B* **72**, 241302 (2005).

[107] Chapon, L. C., Radaelli, P. G., Blake, G. R., Park, S. & Cheong, S.-W. Ferroelectricity induced by acentric spin-density waves in YMn_2O_5. *Phys. Rev. Lett.* **96**, 097601 (2006).

[108] Blake, G. R. *et al.* Spin structure and magnetic frustration in multiferroic RMn_2O_5 (R=Tb,Ho,Dy). *Phys. Rev. B* **71**, 214402 (2005).

[109] Garcia-Flores, A. F. *et al.* Anomalous phonon shifts in the paramagnetic phase of multiferroic RMn_2O_5 (R=Bi, Eu, Dy): Possible manifestations of unconventional magnetic correlations. *Phys. Rev. B* **73**, 104411 (2006).

[110] Valdes Aguilar, R., Sushkov, A. B., Park, S., Cheong, S.-W. & Drew, H. D. Infrared phonon signatures of multiferroicity in $TbMn_2O_5$. *Phys. Rev. B* **74**, 184404 (2006).

[111] Mihailova, B. *et al.* Temperature-dependent raman spectra of $HoMn_2O_5$ and $TbMn_2O_5$. *Phys. Rev. B* **71**, 172301 (2005).

[112] Polyakov, V. *et al.* Coupled magnetic and structural transitions in $EuMn_2O_5$ as studied by neutron diffraction and three-dimensional polarization analysis. *Physica B* **297**, 208–212 (2001).

[113] Kagomiya, I. *et al.* Lattice distortion at ferroelectric transition of YMn_2O_5. *Ferroelectrics* **286**, 167–174 (2003).

[114] Moskvin, A. S. & Pisarev, R. V. Charge-transfer transitions in mixed-valent multiferroic $TbMn_2O_5$. *Phys. Rev. B* **77**, 060102 (2008).

[115] Yamauchi, K., Freimuth, F., Blügel, S. & Picozzi, S. Magnetically induced ferroelectricity in orthorhombic manganites: Microscopic origin and chemical trends. *Phys. Rev. B* **78**, 014403 (2008).

[116] Wang, C., Guo, G.-C. & He, L. First-principles study of the lattice and electronic structure of $TbMn_2O_5$. *Phys. Rev. B* **77**, 134113 (2008).

[117] Giovannetti, G. & van den Brink, J. Electronic correlations decimate the ferroelectric polarization of multiferroic $HoMn_2O_5$. *Phys. Rev. Lett.* **100**, 227603 (2008).

[118] Cullity, B. D. & Graham, C. D. *Introduction to magnetic materials* (Wiley - IEEE Press, 2008).

[119] Hubert, A. & Schäfer, R. *Magnetic domains: The Analysis of Magnetic Microstructures* (Springer, Berlin, 1998).

[120] Jiles, D. C. *Introduction to Magnetism and Magnetic Materials* (Taylor & Francis, 1998).

[121] du Trémolet de Lacheisserie, ., Gignoux, D. & Schlenker, M. (eds.) *Magnetism: Fundamentals* (Springer, 2004).

[122] Palmer, S. B. Antiferromagnetic domains in rare earth metals and alloys. *J. Phys. F* **5**, 2370–2378 (1975).

[123] Palmer, S. B., Baruchel, J., Drillat, A., Patterson, C. & Fort, D. Influence of the purity on the domains and the elastic constants in the helimagnetic phases of Ho and Tb. *J. Magn. Magn. Mater.* **54**, 1626–1628 (1986).

[124] Fiebig, M., Pavlov, V. V. & Pisarev, R. V. Second-harmonic generation as a tool for studying electronic and magnetic strucutres of crystals. *J. Opt. Soc. Am. B* **22**, 96–118 (2005).

[125] Shen, Y. R. *The principles of nonlinear optics* (Wiley & Sons, 2002).

[126] Boyd, R. W. *Nonlinear optics* (Academic Press, 2008).

[127] Pershan, P. S. Nonlinear optical properties of solids: Energy considerations. *Phys. Rev.* **130**, 919–929 (1963).

[128] Sa, D., Valenti, R. & Gros, C. A generalized ginzburg-landau approach to second harmonic generation. *Eur. Phys. J. B* **14**, 301–305 (2000).

[129] Lottermoser, T. *Elektrische und magnetische Ordnung hexagonaler Manganite*. Ph.D. thesis, Universität Dortmund (2002).

[130] Fix, A. *Untersuchung der spektralen Eigenschaften von optischen parametrischen Oszillatoren aus dem optisch nichtlinearen Material Betabariumborat*. Ph.D. thesis, Universität Kaiserslautern (1995).

[131] Loudon, R. *The Quantum Theory of Light* (Oxford Science Publications, 2000).

[132] Fromme, B., Brunokowski, U. & Kisker, E. d-d excitations and interband transitions in mno: A spin-polarized electron-energy-loss study. *Phys. Rev. B* **58**, 9783–9792 (1998).

[133] Fiebig, M., Fröhlich, D., Lottermoser, T. & Maat, M. Probing of ferroelectric surface and bulk domains in $RMnO_3$ (R=Y, Ho) by second harmonic generation. *Phys. Rev. B* **66**, 144102 (2002).

[134] Berkovic, G., Shen, Y. R., Marowsky, G. & Steinhoff, R. Interference of second harmonic generation from a substrate and from an adsorbate layer. *J. Opt. Soc. Am. B* **6**, 205–208 (1989).

[135] Kemnitz, K. *et al.* The phase of second-harmonic light generated in an interface and its relation to absolute molecular orientation. *Chem. Phys. Lett.* **131**, 285–290 (1986).

[136] Fiebig, M., Fröhlich, D., Leute, S. & Pisarev, R. V. Topography of antiferromagnetic domains using second harmonic generation with an external reference. *Appl. Phys. B* **66**, 265–270 (1998).

[137] Becker, P., Bohatý, L., Eichler, H. J., Rhee, H. & Kaminskii, A. A. High-gain raman induced multiple stokes and anti-stokes generation in monoclinic multiferroic $MnWO_4$ single crystals. *Laser Phys. Lett.* **4**, 884–889 (2007).

[138] Momma, K. & Izumi, F. Vesta: A three-dimensional visualization system for electronic and structural analysis. *J. Appl. Crystallogr.* **41**, 653–658 (2008).

[139] Ehrenberg, H., Weitzel, H., Fuess, H. & Hennion, B. Magnon dispersion in $MnWO_4$. *J. Phys.: Condens. Matter* **11**, 2649 – 2659 (1999).

[140] Dachs, H. Zur Deutung der magnetischen Struktur des Hübernits, $MnWO_4$. *Solid State Commun.* **7**, 1015–1017 (1969).

[141] Lautenschläger, G. *et al.* Magnetic phase transitions of $MnWO_4$ studied by the use of neutron diffraction. *Phys. Rev. B* **48**, 6087 – 6098 (1993).

[142] Tian, C. *et al.* Magnetic structure and ferroelectric polarization of $MnWO_4$ investigated by density functional calculations and classical spin analysis. *Phys. Rev. B* **80**, 104426 (2009).

[143] Sagayama, H. *et al.* Correlation between ferroelectric polarization and sense of helical spin order in multiferroic $MnWO_4$. *Phys. Rev. B* **77**, 220407(R) (2008).

[144] Arkenbout, A. H., Palstra, T. T. M., Siegrist, T. & Kimura, T. Ferroelectricity in the cycloidal spiral magnetic phase of $MnWO_4$. *Phys. Rev. B* **74**, 184431 (2006).

[145] Mitamura, H., Nakamura, H., Kimura, T., Sakakibara, T. & Kindo, K. Sign reversal of the dielectric polarization of $MnWO_4$ in very high magnetic fields. *J. Phys.: Conf. Ser.* **150**, 042126 (2009).

[146] Harris, A. B. Landau analysis of the symmetry of the magnetic structure and magnetoelectric interaction in multiferroics. *Phys. Rev. B* **76**, 054447 (2007).

[147] Tolédano, P., Mettout, B., Schranz, W. & Krexner, G. Directional magnetoelectric effects in $MnWO_4$: Reality of the Dzyaloshinskii-Moriya interaction in magnetic multiferroics. *unpublished* **00**, 00 (2009).

[148] Maringer, M. *Spektroskopie und Domänentopographie an multiferroischem $MnWO_4$ mittels nichtlinearer Optik.* Master's thesis, Universität Bonn (2008).

[149] Han, T.-C. & Lin, J. G. Effect of iron doping on the magnetic properties of $TbMn_2O_5$. *J. Magn. Magn. Mater.* **310**, 355–357 (2007).

[150] Bertaut, E. F. *et al.* Compounds of rare earth oxides with transition metal oxides. *Bull. Soc. Chim. Fr.* **4**, 1132–1137 (1965).

[151] Abrahams, S. C. & Bernstein, J. L. Crystal structure of paramagnetic $DyMn_2O_5$ at 298 k. *J. Chem. Phys.* **46**, 3776–3782 (1967).

[152] Noda, Y. *et al.* Magnetic and ferroelectric properties of multiferroic RMn_2O_5. *J. Phys.: Condens. Matter* **20**, 434206 (2008).

[153] Kimura, H., Noda, Y. & Kohn, K. Spin-driven ferroelectricity in the multiferroic compounds of RMn_2O_5. *J. Magn. Magn. Mater.* **321**, 854–857 (2009).

[154] Kobayashi, S. *et al.* Neutron diffraction study of successive magnetic phase transitions in ferroelectric $TbMn_2O_5$. *J. Phys. Soc. Jpn.* **73**, 3439–3443 (2004).

[155] Wang, C., Guo, G.-C. & He, L. Ferroelectricity driven by the noncentrosymmetric magnetic ordering in multiferroic $TbMn_2O_5$: A first-principles stud. *Phys. Rev. Lett.* **99**, 177?0? (2007).

[156] Radaelli, P. G. & Chapon, L. C. A neutron diffraction study of RMn_2O_5 multiferroics. *J. Phys.: Condens. Matter* **20**, 434213 (2008).

[157] Koo, J. *et al.* Non-resonant and resonant X-ray scattering studies on multiferroic $TbMn_2O_5$. *Phys. Rev. Lett.* **99**, 197601 (2007).

[158] Nogami, A., Suzuki, T. & Katsufuji, T. Second harmonic generation from multiferroic $MnWO_4$. *J. Phys. Soc. Jpn.* **77**, 115001 (2008).

[159] Ejima, T. *et al.* Microscopic optical and photoelectron measurements of MWO_4 (m = mn, fe and ni). *J. Lumin.* **119 - 120**, 59 – 63 (2006).

[160] Huffmann, D. R., Wild, R. L. & Shinmei, M. Optical absorption spectra of crystal-field transitions in MnO. *J. Chem. Phys.* **50**, 4092 (1969).

[161] Wood, E. A. *Crystals and light: An in introduction to optical crystallography* (Dover Publications, 1977).

[162] Li, F. H. & Cheng, Y. F. A simple approach to quasicrystal structure and phason defect formulation. *Acta Crystallogr., Sect. A* **A46**, 142–149 (1990).

[163] Lawes, G., Kenzelmann, M. & Broholm. Magnetically induced ferroelectricity in the buckeld kagome antiferromagnet $Ni_3V_2O_8$. *J. Phys.: Condens. Matter* **20**, 434205 (2008).

[164] Meier, D., Aliouane, N., Argyriou, D. N., Mydosh, J. A. & Lorenz, T. New features in the phase diagram of $TbMnO_3$. *New J. Phys.* **9**, 100 (2007).

[165] Baier, J. *et al.* Hysteresis effects in the phase diagram of multiferroic $GdMnO_3$. *Phys. Rev. B* **73**, 100402(R) (2006).

[166] Chaudhury, R. P. *et al.* Thermal expansion and pressure effect in $MnWO_4$. *Physica B* **403**, 1428–1430 (2007).

[167] Prokhnenko, O. *et al.* Enhanced ferroelectric polarization by induced Dy spin order in multiferroic $DyMnO_3$. *Phys. Rev. Lett.* **98**, 057206 (2006).

[168] Fennie, C. J. & Rabe, K. M. Ferroelectric transition in $YMnO_3$ from first principles. *Phys. Rev. B* **72**, 100103 (2005).

[169] Palai, R. *et al.* β phase and $\gamma - \beta$ metal-insulator transition in multiferroic $BiFeO_3$. *Phys. Rev. B* **77**, 014110 (2008).

[170] Lebeugle, D., D. andColson, Forget, A. & Viret, M. Very large spontaneous electric polarization in $BiFeO_3$ single crystals at room temperature and its evolution under cycling fields. *Appl. Phys. Lett.* **91**, 022907 (2007).

[171] Camlibel, I. Spontaneous polarization measurements in several ferroelectric oxides using a pulsed-field method. *Appl. Phys. Lett.* **40**, 1690 (1969).

[172] Seul, M. & Andelman, D. Domain shapes and patterns: The phenomenology of modulated phases. *Science* **267**, 476–483 (1995).

[173] Evans, P. G., Isaacs, E. D., Aeppli, G., Cai, Z. & Lai, B. X-ray microdiffraction images of antiferromagnetic domain evolution in chromium. *Science* **295**, 1042–1045 (2002).

[174] Li, Y. Y. Domain walls in antiferromagnets and the weak ferromagnetism of α-Fe_2O_3. *Phys. Rev.* **101**, 1450–1454 (1956).

[175] Finger, T. *et al.* Control of chiral magnetism in multiferroic $MnWO_4$ through an electric field. *arXiv* 0907.5319 (2009).

[176] Terhune, R. W., Maker, P. D. & Savage, C. M. Optical harmonic generation in calcite. *Phys. Rev. Lett.* **8**, 404–406 (1962).

[177] Lee, C. H., Chang, R. K. & Bloembergen, N. Nonlinear electroreflectance in silicon and silver. *Phys. Rev. Lett.* **18**, 167–170 (1967).

[178] Lee, S., Ratcliff, W., Cheong, S.-W. & V., K. Electric field control of the magnetic state in bifeo$_3$ single crystals. *Appl. Phys. Lett.* **92**, 192906 (2008).

[179] Lee, S. *et al.* Single ferroelectric and chiral magnetic domain of single-crystalline bifeo$_3$ in an electric field. *Phys. Rev. B* **78**, 100101(R) (2008).

[180] Volk, T. & Wöhlecke, M. *Lithium Niobate: Defects, Photorefraction and Ferroelectric Switching* (Springer-Verlag, 2008).

[181] Böttger, U. & Summerfelt, S. R. *Nanoelectronics and Information Technology: Advanced Electronic Materials and Novel Devices* (Wiley-VCH Verlag, 2005).

[182] Kittel, C. *Einführung in die Festkörperphysik* (Oldenbourg Wissenschaftsverlag, 2002).

[183] Kordel, T. *et al.* Nanodomains in multiferroic hexagonal $RMnO_3$ films (R=Y,Dy,Ho,Er). *Phys. Rev. B* **80**, 045409 (2009).

[184] Beaurepaire, E., Merle, J.-C., Daunois, A. & Bigot, J.-Y. Ultrafast spin dynamics in ferromagnetic nickel. *Phys. Rev. Lett.* **76**, 4250–4254 (1996).

[185] Armstrong, J. A., Bloembergen, N., Ducuing, J. & Pershan, P. S. Interactions between light waves in a nonlinear dielectric. *Phys. Rev.* **127**, 1918–1939 (1962).

[186] Maker, P. D., Terhune, R. W., Nisenoff, M. & Savage, C. M. Effects of dispersion and focusing on the production of optical harmonics. *Phys. Rev. Lett.* **8**, 21–22 (1962).

Awards and Publications

Awards

- Opto-Electronics Committee Prize; Awarded for the Best Contributed Paper at the Mini-Symposium *Periodically-Modulated and Artificially Hetero-Structured Devices*, Grasmere, May 18 – 21, 2009

- MRS Graduate Student Award finalist for the 2009 MRS Fall Meeting; The distinction between the Gold and Silver Awards will be made during the MRS Fall Meeting, Boston, November 30 – December 1, 2009

Publications

Articles in referred journals

(10) *Imaging of Hybrid-Multiferroic and Translation Domains in a Spin-Spiral Ferroelectric*, D. Meier, N. Leo, M. Maringer, Th. Lottermoser, P. Becker, L. Bohatý, M. Fiebig, to be published in the MRS Proceedings (2009)

(9) *Topology and manipulation of multiferroic hybrid domains in $MnWO_4$*, D. Meier, N. Leo, M. Maringer, Th. Lottermoser, P. Becker, L. Bohatý, M. Fiebig, accepted for publication in Phys. Rev. B (2009)

(8) *Giant coupling of second-harmonic generation to a multiferroic polarization*, Th. Lottermoser, D. Meier, R. V. Pisarev, and M. Fiebig, Phys. Rev. B 80, 100101(R) (2009)

(7) *Nanodomains in multiferroic hexagonal $RMnO_3$ films (R = Y, Dy, Ho, Er)*, T. Kordel, C. Wehrenfennig, D. Meier, Th. Lottermoser, and M. Fiebig, I. Gélard and C. Dubourdieu, J.-W. Kim, L. Schultz, and K. Dörr, Phys. Rev. B 80, 045409 (2009)

(6) *Observation and coupling of domains in a spin-spiral multiferroic*, D. Meier, M. Maringer, Th. Lottermoser, P. Becker, L. Bohatý, M. Fiebig, Phys. Rev. Lett. 102, 107202 (2009)

(5) *Anomalous thermal expansion and strong damping of the thermal conductivity of $NdMnO_3$ and $TbMnO_3$ due to 4f crystal-field excitations*, K. Berggold, J. Baier, D. Meier, J. A. Mydosh, T. Lorenz, J. Hemberger, A. Balbashov, N. Aliouane, D. N. Argyriou, Phys. Rev. B 76, 094418 (2007)

(4) *New features in the phase diagram of $TbMnO_3$*, D. Meier, N. Aliouane, D. N. Argyriou, J. A. Mydosh, T. Lorenz, NJP 9, 100, (2007)

(3) *Uniaxial pressure dependencies of the phase transitions in GdMnO₃*, J. Baier, D. Meier, K. Berggold, J. Hemberger, A. Balbashov, J. A. Mydosh, T. Lorenz, J. Magn. Magn. Mat. 310, 1165 (2007)

(2) *Hysteresis effects in the phase diagram of multiferroic GdMnO₃*, J. Baier, D. Meier, K. Berggold, J. Hemberger, A. Balbashov, J. A. Mydosh, T. Lorenz, Phys. Rev. B 73, 100402(R) (2006)

(1) *Nature of Low Temperature Phase Transitions in $CaMn_7O_{12}$*, O. Volkova, Y. Arango, N. Tristan, V. Kataev, E. Gudilin, D. Meier, T. Lorenz, B. Büchner, A. Vasiliev, JETP Lett. 82, 444 (2005)

Conference contributions – TALKS

(12) ISFD-10 Conference, Prag, September 20-24, 2010: *Imaging and control of domains in a spin-spiral multiferroic*, D. Meier and M. Fiebig

(11) DPG Conference, Regensburg, March 21-26, 2010: *Topological magnetoelectric memory effect in the spin-spiral multiferroic $MnWO_4$*, D. Meier, N. Leo, Th. Lottermoser, P. Becker, L. Bohatý, M. Fiebig

(10) MRS Fall Meeting (Special Talk Sessions, MRS Graduate Student Award), Boston, November 30 - December 4, 2009: *Domain topology and switching behaviour of magnetically-induced ferroelectrics*, D. Meier, N. Leo, M. Maringer, Th. Lottermoser, P. Becker, L. Bohatý, M. Fiebig

(9) MRS Fall Meeting, Boston, November 30 - December 4, 2009: *Switching behaviour of magnetically induced ferroelectric domains in $MnWO_4$*, D. Meier, N. Leo, M. Maringer, Th. Lottermoser, P. Becker, L. Bohatý, M. Fiebig

(8) Mini-Symposium on Periodically-Modulated and Artificially Hetero-Structured Devices, Grasmere, 18-21 May, 2009: *Control of ferroelectric and antiferromagnetic domains in a spin-spiral multiferroic*, D. Meier, N. Leo, M. Maringer, Th. Lottermoser, P. Becker, L. Bohatý, M. Fiebig

(7) DPG Conference, Dresden, March 22-27, 2009: *Controlled manipulation and coupling of domains in a spin spiral multiferroic*, D. Meier, M. Maringer, N. Leo, Th. Lottermoser, P. Becker, L. Bohatý, M. Fiebig

(6) March Meeting of the APS, Pittsburgh, March 16-20, 2009: *Observation and coupling of domains in a spin spiral multiferroic*, D. Meier, M. Maringer, Th. Lottermoser, P. Becker, L. Bohatý, M. Fiebig

(5) PhD-Colloquium of SFB 608, Cologne, July 9, 2008: *Observation and control of antiferromagnetic and ferroelectric domains in $MnWO_4$*, D. Meier, M. Maringer, Th. Lottermoser, P. Becker, L. Bohatý, M. Fiebig

(4) March Meeting of the APS, New Orleans, March 10-14, 2008: *Ferroelectric domain topology of the multiferroic spin spiral system* $MnWO_4$, D. Meier, Th. Lottermoser, P. Becker, L. Bohatý, M. Fiebig

(3) DPG Conference, Berlin, February 25-29, 2008: *Coexistence of ferroelectric and long-wavelength magnetic ordering in* $MnWO_4$, D. Meier, M. Maringer, Th. Lottermoser, P. Becker, L. Bohatý, M. Fiebig

(2) MaCoMuFi Meeting, Bonn, November 07-09, 2007: *Advanced characterization of multiferroic* $RMnO_3$ *films by Second Harmonic Generation*, D. Meier, T. Kordel, C. Wehrenfennig, Th. Lottermoser, M. Fiebig

(1) DPG Conference, Regensburg, March 26-30, 2007: *New features in the phase diagram of* $TbMnO_3$, D. Meier, N. Aliouane, D. Argyriou, J. A. Mydosh, T. Lorenz

Conference contributions – POSTERS

(26) DPG Conference, Regensburg, March 21-26, 2010: *Separation and magnetic-field dependence of contributions to the magnetically induced net polarization in multiferroic* $TbMn_2O_5$, N. Leo, Th. Lottermoser, D. Meier, R. V. Pisarev, M. Fiebig

(25) DPG Conference, Regensburg, March 21-26, 2010: *Switching of a spin-spiral-induced polarization in multiferroic* $MnWO_4$, T. Hoffmann, D. Meier, P. Becker, L. Bohatý, M. Fiebig

(24) DPG Conference, Regensburg, March 21-26, 2010: *Advanced imaging and characterization of ferroic domains by optical second harmonic generation*, D. Meier, C. Wehrenfennig, M. Fiebig

(23) SFB 608 Regional Workshop on Correlated Systems 2009, Kerkrade, September 30 - October 2, 2009:*Switching behavior of multiferroic domains in* $MnWO_4$, D. Meier, T. Hoffmann, N. Leo, M. Maringer, Th. Lottermoser, P. Becker, L. Bohatý, M. Fiebig

(22) SFB 608 Regional Workshop on Correlated Systems 2009, Kerkrade, September 30 - October 2, 2009: *Composite nature of magnetically induced spontaneous polarization in* $TbMn_2O_5$, N. Leo, Th. Lottermoser, D. Meier, R. Pisarev, M. Fiebig

(21) ESMF-3 2009, European School on Multiferroics, Groningen, September 7-11, 2009: *Nonlinear optical spectroscopy on* $TbMn_2O_5$, N. Leo, Th. Lottermoser, D. Meier, R. Pisarev, M. Fiebig

(20) ESMF-3 2009, European School on Multiferroics, Groningen, September 7-11, 2009: *Switching behavior of multiferroic domains in* $MnWO_4$, T. Hoffmann, M. Maringer, D. Meier, Th. Lottermoser, P. Becker, L. Bohatý, M. Fiebig

(19) BCGS Poster Session 2009, Bonn, June 9, 2009: *Ultrafast switching of a multiferroic polarization*, T. Hoffmann, M. Maringer, D. Meier, Th. Lottermoser, P. Becker, L. Bohatý, M. Fiebig

(18) DPG Conference, Dresden, March 22-27, 2009: *Topology of the electric order in multiferroic orthorhombic DyMnO$_3$*, T. Günter, D. Meier, Th. Lottermoser, D. Argyriou, M. Fiebig

(17) DPG Conference, Dresden, March 22-27, 2009: *Analysis of ferroelectric and magnetic chiral order in MnWO$_4$*, N. Leo, D. Meier, Th. Lottermoser, M. Maringer, P. Becker, L. Bohatý, M. Fiebig

(16) Workshop on Magnetoelectric Interaction Phenomena in Crystals (MEIPIC6), Santa Barbara, January 25-28, 2009: *Nonlinear optical spectroscopy on mulitferroic TbMn$_2$O$_5$*, Th. Lottermoser, D. Meier, A. Zimmermann, R. V. Pisarev, M. Fiebig

(15) ICMR Multiferroics and beyond Summer Program, Santa Barbara, July 20-August 2, 2008: *Investigation of multiferroic spin spiral systems by Second Harmonic Generation*, D. Meier, M. Maringer, Th. Lottermoser, P. Becker, L. Bohatý, M. Fiebig

(14) E-MRS 2008 Spring Meeting, Strasbourg, May 26-30, 2008: *Characterization of multiferroic HoMnO$_3$ films by optical second harmonic generation*, T. Kordel, C. Wehrenfennig, D. Meier, Th. Lottermoser, I. Gerlard, C. Dubourdieu, M. Fiebig

(13) DPG Conference, Berlin, February 25-29, 2008: *Investigation of multiferroic properties in MnWO$_4$ by SHG-spectroscopy*, M. Maringer, D. Meier, Th. Lottermoser, P. Becker, L. Bohatý, M. Fiebig

(12) DPG Conference, Berlin, February 25-29, 2008: *Characterization of multiferroic HoMnO$_3$ films by second harmonic generation*, T. Kordel, C. Wehrenfennig, D. Meier, Th. Lottermoser, C. Dubourdieu, I. Gerald, K. Doerr, J.-W. Kim, M. Fiebig

(11) SFB 608 Workshop on Quantum Matter, Köln, October 10-12, 2007: *Coexistence of magnetic and electric domains in multiferroic MnWO$_4$*, D. Meier, G. Yuan, Th. Lottermoser, P. Becker, L. Bohatý, M. Fiebig

(10) SFB 608 Workshop on Quantum Matter, Köln, October 10-12, 2007: *Enhanced multiferroic properties in HoMnO$_3$*, T. Kordel, D. Meier, Th. Lottermoser, K. Doerr, C. Dubourdieu, M. Fiebig

(9) DPG Conference, Regensburg,March 26-30, 2007: *Thermal Expansion in NdMnO$_3$ and TbMnO$_3$*, K. Berggold, T. Lorenz, J. Baier, D. Meier, J. Hemberger, D. Argyriou, J. A. Mydosh

(8) SFB 608 Workshop on Strongly Correlated Transition Metal Compounds II, Köln, September 11-14, 2006: *Magnetoelastic coupling of multiferroic manganites*, J. Rohrkamp, J. Baier, D. Meier, K. Berggold, O. Heyer, J. Hemberger, A. Balbashov, N. Aliouane, D. Argyriou, A. Freimuth, J. A. Mydosh, T. Lorenz

(7) International Conference on Magnetism, Kyoto, August 20 - August 25, 2006: *Thermal properties of multiferroic RMnO$_3$ (R = Gd, Tb, and Dy)*, J. Baier, D. Meier, K. Berggold,

J. Hemberger, A. Balbashov, N. Aliouane, D. N. Argyriou, J. A. Mydosh, T. Lorenz

(6) EPS / DPG Conference on Magnetic Properties, Dresden, March 26-31, 2006: *Thermal expansion of multiferroic manganites in magnetic fields*, D. Meier, J. Baier, O. Heyer, J. Hemberger, D. Argyriou, N. Aliouane, A. Freimuth, T. Lorenz

(5) EPS / DPG Conference on Magnetic Properties, Dresden, March 26-31, 2006: *Thermal transport in multiferroics*, K. Berggold, T. Lorenz, J. Baier, D. Meier, J. Hemberger, D. Argyriou, A. Vasiliev

(4) EPS / DPG Conference on Magnetic Properties, Dresden, March 26-31, 2006: *Thermodynamic and transport properties of doped La_2CoO_4*, E. Rose, N. Hollmann, J. Baier, M. Benomar, K. Berggold, M. Cwik, M. Kriener, D. Meier, A. Tanaka, T. Lorenz, A. Freimuth

(3) EPS / DPG Conference on Magnetic Properties, Dresden, March 26-31, 2006: *Phase diagram of the multiferroic $GdMnO_3$ studied by thermal expansion and magnetostriction*, J. Baier, D. Meier, V. Ivanov, A. Mukhin, A. Balbashov, J. Hemberger, T. Lorenz

(2) Symposium on Functional Transition Metal Compounds and Multiferroics, Köln, September 26-28, 2005: *Thermal Expansion and Transport of Multiferroic Manganites*, D. Meier, K. Berggold, J. Baier, O. Heyer, J. Hemberger, A. Vasiliev, V.Y. Ivanov, A.A. Mukhin, A.M. Balbashov, D. Argyriou, N. Aliouane, A. Freimuth, T. Lorenz

(1) Symposium on Functional Transition Metal Compounds and Multiferroics, Köln, September 26-28, 2005: *Crystal structure and magnetic properties of $La_{2-x}Sr_xCoO_4$*, M. Cwik, M. Benomar, M. Reuther, M. Haider, A. Hoser, Y. Sidis, E. Rose, D. Meier, J. Baier, T. Lorenz, M. Braden